CW00798632

To Michael Fletcher
from Hiram Baddeley
May 2009.

Physics and the Human Body

Stories of Who Discovered What

by

Hiram Baddeley

AuthorHouse™ UK Ltd.
500 Avebury Boulevard
Central Milton Keynes, MK9 2BE
www.authorhouse.co.uk
Phone: 08001974150

First published by AuthorHouse 11/17/2008

ISBN: 978-1-4389-1703-0 (sc)

Printed in the United States of America
Bloomington, Indiana

This book is printed on acid-free paper.

FOR DAMIAN

CONTENTS

ABBREVIATIONS

AC	alternating current
ADP	adenosine diphosphate
AF	atrial fibrillation
ATP	adenosine triphosphate
BMR	basal metabolic rate
BP	arterial blood pressure
c	velocity of light
CAT	computerized axial (x-ray) tomography
cgs	*centimetre, gram, second* system of measures
CPR	cardiopulmonary resuscitation
CRO	cathode ray oscilloscope
csf	cerebrospinal fluid
CT	computer-assisted (x-ray) tomography
CXR	chest radiograph
DC	direct current
DNA	deoxy-ribose nucleic acid
ECG	electrocardiogram
EEG	electroencephalogram
emf	electromotive force
ERCP	endoscopic retrograde cholangiopancreatography

h	Planck's constant
IVC	inferior vena cava
iPO_4	inorganic phosphate
LA	left atrium
LV	left ventricle
MRI	magnetic resonance imaging
NMR	nuclear magnetic resonance
PA	pulmonary artery
Pco_2	partial pressure of carbon dioxide
Po_2	partial pressure of oxygen
pd	potential difference
PET	positron emission tomography
RA	right atrium
RF	radiofrequency
RNA	ribose nucleic acid
RV	right ventricle
SI	*Système International* of weights and measures
SPECT	single photon emission computed tomography
UV	ultraviolet radiation
VF	ventricular fibrillation
XR	x-ray radiograph

NOTES and READING

More detailed information about technical topics is contained in the NOTES section at the end of each chapter, referenced by superscript numbers in the text.

The sources of information and opinions are derived from the chapters and books listed in the READING section at the end of the book.

PREFACE

Scientific theories about what we are and how our bodies function; what the world around us is and how the laws of nature work, have concerned natural philosophers and physicians for several millennia. Until the 19th century, understanding of what we now call physics, developed alongside the understanding of human biology. Before then all science had been included within the vague but universal domain of *natural philosophy*. The history of ideas in physics and biology was that of informal collaboration between them.

Then divisions were made in the 1830s when the Cambridge polymath, Master of Trinity, and philosopher of science, William Whewell (1794 – 1866), confined the study of *mechanics, heat, optics, astronomy, electricity* and *magnetism* " *within the jurisdiction of mathematics*" to comprise the '*exact*' science of *physics*, which was to be distinct from *biology* and *chemistry*. He gave a lecture on the philosophy of science to the British Association for the Advancement of Science in 1833 and used the term '*scientist*' for the first time. After this time '*hard physics*' and '*soft biology*' developed as separate scientific cultures. The distinctions between the three branches of science became accepted and have persisted until quite recently.

A physicist, Ernest Rutherford (1871 – 1937), boasted that "*All science is either physics or stamp collecting*", which implied that science is either mathematical and quantitative, or descriptive. His scientific research initiated the process of unifying *chemistry* and *atomic physics* for which he received the Nobel Prize for Chemistry. But for Rutherford *biology*, including *human biology* was still just '*stamp collecting*'.

This book affirms that the separation between *physics* and *human biology* is artificial. The laws of physics apply at every level to the molecules, cells, tissues and organs of our bodies, but many people are bewildered because physics and biology have been presented as separate sciences. The purpose of this book is to show historically how discoveries in physics and human biology have been related to one another, by cross-fertilization of ideas and collaboration between physicists, physicians, medical scientists and engineers.

Descriptions of the science have been combined with brief biographies of the scientists, to give a coherent history of theories and discoveries. Theories have been explained as they developed so that the science is not separated from the scientist, nor is physics separated from medicine. *Cardiac, respiratory* and *neuromuscular physiology* and the *discovery of DNA* are interleaved with explanations of *Newton's Laws, Faraday's Electricity, Laws of Thermodynamics, Maxwell's electromagnetic radiation* and *Quantum Theory.*

It is easiest to follow the related stories by grouping discoveries under the headings of the major branches of physics:

mechanics with *human motion*
fluid mechanics with *blood flow* and *acoustics*
thermodynamics with *metabolism*
electricity with *nerve and muscle function*
light with *vision*
atomic theory with *molecular biology*

Some conjectures about the processes of discovery and the motivation of individuals have been included. For example; the monk Sarpi of Venice inspired both Galileo and William Harvey. The young Oxford physiologists – Wren, Hooke, and Lower, became obsessed by blood and brains after they were locked in Westminster School as Charles I was beheaded in nearby Whitehall.

Einstein realized that quantum events in the retina linked cosmology and particle physics. The statistical nature of quantum physics led Schrödinger, a physicist and philosopher, to write an essay entitled *What is life?* By applying the principles of *quantum mechanics* he deduced that the genetic component of chromosomes, the '*Key of Life*', must be a very stable *aperiodic crystal*. Schrödinger's essay inspired two physicists, an x-ray crystallographer and a geneticist to discover the structure and role of deoxyribose nucleic acid (DNA).

I hope this book will be attractive to science and medical students of all ages, those readers who are interested in the history of ideas and who want to know how we came to understand the way our bodies function, according to the laws of physics.

I would like to express my appreciation to all my colleagues; physicists, radiographers or physicians, who have contributed to my understanding of medicine, clinical radiology, human biology and physics over many years. In particular Oscar Craig and David Sutton at St Mary's Hospital, London; Rhys Davies, John Roylance, Howard Middlemiss, Peter Wells, Glenda Bryan and Frank Ross at the United Bristol Hospitals and University of Bristol; Wai Lup Wong, Michele Saunders, Stan Dische, Linda Culver, Peter Dovey, Edwin Aird, Gary Glover and Roy Sanders at Mount Vernon Hospital, Northwood; Tom Sherwood and Adrian Dixon, New Addenbrook's Hospital and Cambridge University; Lawrie Powell, Graham Cooksley, John Partridge, Mark Benson, Jim Williams, Cam Battersby, David Dodderell, John Earwaker and the Faculty of Medicine in the University of Queensland; Brian Thomas and Brian Thomas, Physics Department, Queensland University of Technology; also my fellow radiologists in the Royal Colleges of Radiology in Australasia and the United Kingdom.

The librarians at the Whipple Museum in Cambridge, the Royal Society of Medicine and Wellcome Libraries in London, and the Library of Western Australia were very helpful in finding appropriate background and reference texts.

I am grateful to all the members of my family for their support, and to friends who read sections and made useful suggestions; especially to Mike McGrath and Chris Bradley, who weeded out some of my worst errors; their help is greatly appreciated.

Hiram Baddeley
Somerset 2008.

PHILOSOPHERS, PHYSICISTS & PHYSICIANS

Most of us live quite adequately without needing to know how our bodies work. Like Voltaire's ladies of Paris we can *"live well without knowing what goes into the stew"*. In health we breathe, eat and sleep almost instinctively, only when sick do we worry about our *insides*'. Living our bodies in their surroundings is physical. We see light, hear sound, feel heat, use force to either lift weight or move ourselves.

The understanding of the form and function of the human body has been intertwined with that of the material world over the last three millennia. The ancient Greek term for nature – *phusis-* is the root for both physics and physician. There are close links between the discoveries of physics, biology and medicine.

How did we come to know as much as we do about the physical world around us, our own bodies and how they function? The common history of medicine and physics is littered with speculations that seemed plausible initially but which ultimately failed to demonstrate credibility, such as the theories of *humours, pneuma, phlogiston, ether* and *homocentric (isocentric) astronomy*. Others, such as *Copernican astronomy, Kepler's laws of planetary motion, Newton's laws* and *Harvey's circulation of the blood* provided durable hypotheses, which though superseded, provided strong foundations for improved theories.

The biographies of thinkers, especially those of the ancient world, are fragmentary, derived from secondary sources, translations and myths. Context and detail are both essential to history, which is the essence of innumerable biographies. Without some biographical detail history is just propaganda and without context history has no relevance

The philosophers and physicists in this history were usually mathematicians and sometimes astronomers, engineers or chemists. The physicians were sometimes anatomists, surgeons, physiologists or microscopists. Galileo trained as both a physician and mathematician before embarking on experimental physics and astronomy. Philosopher/physicists and physicians sometimes worked alongside; the link between Democritus and Hippocrates may be apocryphal, but Galen and Marcus Aurelius, Galileo and Fabricius, Borelli and Malpighi, Newton and Locke, Marie Curie and Claudius Regaud, Hodgkin and Huxley communicated frequently. Some like Gilbert, Stensen, Mayow, Stahl, Black, Poiseuille, Young, Helmholtz , Berzelius and Koch were both trained physicists and physicians.

Perhaps the ultimate integration of discoveries in physics and medicine was represented by von Helmholtz, the Prussian army surgeon and Director of the Institute of Physics in Berlin who proposed the *First Law of Thermodynamics,* invented the ophthalmoscope and modern physiological optics. Max Planck and Heinrich Hertz were his students and research associates.

In the past, philosophers and physicists tended to be either aristocratic or scholastic. The aristocrats generally had time and resources to think and experiment, but ran risks. Plato narrowly escaped death; Aristotle, Einstein and Schrödinger were exiled; Galileo and Volta imprisoned; Archimedes murdered and Lavoisier executed. The scholars were safe to study and think in the sheltered workshops of their universities so long as they avoided accusations of heresy, which Newton managed but Bruno didn't.

By comparison physicians came from humbler origins and associated with the ruling classes only when their reputations had been established. Their lives were safer. Even tyrants needed to keep a good physician around. All the noted physicists and physicians before 1900 were men, with the exceptions of the aristocrat and physicist Emilie de Chatelet, a friend of Voltaire and strong advocate of Newtonian physics in France, and Marie Curie.

DEFINITIONS

(New Shorter Oxford English Dictionary)

Physics is defined as *"that branch of science that deals with the nature and properties of matter and energy, insofar as they are not dealt with by chemistry and biology; the science whose subject matter includes mechanics, gravitation, heat, sound, light and other radiation, electricity, magnetism and the structure of atoms."*

The exclusion of chemistry and biology, as with most attempts to define boundaries is arbitrary; chemical reactions emit or absorb; heat, light and electricity and these energies impinge on biological systems. The overlap of the three major branches of science is confirmed by the intermediate disciplines of physical chemistry, biomechanics and molecular biology.

The enormous contributions of pharmaceutical chemistry and molecular biology to medicine are generally beyond the scope of this story as also are the extremes of *'Big Science'* physics –*'Big Bang Theory'*, *'Black Holes'*, *quarks* and *'Super string Theory'*. Physiological chemistry is included where necessary but the major purpose of this book is to pursue the historic links between physics, human biology and medicine.

Body has several meanings. The most relevant to this book are

1. *"The material and physical frame of a human or animal; the whole material organism."*
2. *"A separate piece or portion of matter; a material thing of three dimensions; that occupies space. A compact amount; a physical bulk."*

Anatomy is *"the science of the structure of the bodies of humans, animals and plants. It involves the artificial separation of the parts of a human, animal or vegetable body, by dissection"'*- mostly dead bodies.

Physiology on the other hand is *"that branch of science that deals with normal functioning of living organisms and their parts, in so far as it is not dealt with by biochemistry or immunology."* The exclusions mirror those in the definition of physics.

The different definitions of anatomy and physiology give a clue to the dualistic way in which we regard the form and function of our bodies. The definitions sound so different yet they relate two fundamental aspects of the same living bodies. To study morphology without regard to function is scientifically irrelevant; physiology is impossible in the absence of living anatomy.

Medical professionals tend to be blasé about the basic medical sciences of anatomy and physiology, but to many ordinary people the techniques associated with them; of human dissection and animal vivisection, are repugnant. In Confucian China, human dissection was not recognized. Even after the communist Cultural Revolution, medical dissection of cadavers was not approved until the 1980s, when corpses of executed convicts and late-term foetuses became available to medical colleges. Public dissection of cadavers was not encouraged in ancient Greece or Rome, except in Alexandria of the Ptolemys.

From the 14th to early 19th centuries in Europe, horrific public execution rituals made dissection of dead corpses seem a comparatively innocuous activity. In 1539 judges in Padua scheduled executions to fit Vesalius's anatomy teaching timetable. After this time, systematic anatomical dissection became an essential part of medical education and an initiation rite for medical students, though still repugnant to the general public. Even in 21st century Britain there is squeamishness about dissection of cadavers. Not everyone carries an organ donor card. Animal vivisection for physiological or pharmacological research causes outrage, not only amongst animal rights activists. All sane people oppose unnecessary cruelty to animals; the conflict is about what is necessary and what is cruel.

Finally it is necessary to clarify the different usages of **medicine** and **physician** that can be confusing. **Medicine** in its most embracing sense comprises all the disciplines and includes surgery, obstetrics, pathology and psychiatry as well as internal medicine, previously known as *physick*. **Medicine** in its narrow meaning is *internal medicine* and implies non-surgical, usually pharmacological treatments of disease. In Europe the differences between the academic doctors of *physick* in the universities and barber-surgeon tradesmen separated the disciplines of *internal medicine* and *surgery*.

Academic doctors studied in college rooms and botanic gardens whilst surgeons learned their craft in the market place or on the battlefield. Until the mid 19th century internal physicians and surgeons divided the human body between them. The **internists** commanded the **vital organs** contained within the **three major body cavities**: the brain and cord within the **cranial cavity and spinal canal**, lined by meningeal membranes; the heart, great vessels and lungs within the **thorax**, lined by pleural membranes; and the liver, spleen, kidneys and intestines contained within the peritoneal lining of the **abdominal cavity**. Bleedings, purging, fasts and specifics were used to maintain the balance between these vital organs so as to repel disease. Surgeons entered these cavities at their patients' peril; instead they addressed the more accessible parts of the body using their knives, lancets, cauteries and saws on the outer head and trunk, and the limbs. The pelvic contents are not strictly within the peritoneal cavity and so the internist physicians were content to leave the bladder, rectum and sexual organs to surgeons and midwives; however they the examined urine and faeces of their patients most carefully before making prognostications.

Since the mid 19th century, when aseptic surgery and anaesthesia allowed surgeons to enter the abdominal cavity successfully, British surgeons have been required to have diplomas or degrees in medicine, including *internal medicine*; yet the inverted prestige of *"Mister, Miss or Missus"* for surgical specialists continues to confuse their patients. Elsewhere in the world surgeons are known as doctors of medicine and are regarded as physicians who operate, as did Galileo's physician, Fabricius de Aquapedente.

Hippocrates regarded himself as a physician in this larger sense. His last aphorism was – *"What drugs will not cure, the knife will; what the knife will not cure, the cautery will; what the cautery will not cure must be considered incurable."* The medicine of ancient Greece was an art rather than a science but embraced all aspects of treating the sick.

THE GROWTH OF SCIENCE

Philosophy and science have not usually prospered under absolutist regimes whose rulers and enforcers require obedience from their subjects and who jealously reserve the right to interpret the cosmos, the nature of the world and of mankind to themselves. The Chinese emperors and mandarins suppressed innovations in science and some in technology as being a threat to the Mandate of Heaven. Although the Chinese invented clocks, paper, printing, gunpowder and sophisticated systems of astronomy and traditional medicine, these technologies were not developed. The role of the Chinese scientist was not to understand or explain natural phenomena but to adapt them to the service of the Emperor. Chinese medicine was effective but remained empirical. In a similar manner the Christian Roman emperors, popes and bishops required obedience to God and to themselves as His vicars on earth. The cosmology of homocentric spheres was essential to the dogma of celestial Heaven and infernal Hell. To support Copernicus was to deny this dogma. For doing so, the monk Bruno was burned at the stake in Rome in 1600.

To Greece in general and Athens in particular we owe *democracy* – government by open discussion, not royal decree; *philosophy* – rational argument about ultimate questions; *mathematics* – a deductive science and not just rules-of-thumb, and *physics* – provable hypotheses regarding the material world as opposed to myths of world formation. Hellenic culture was based on reason. Although the Egyptians and Babylonians had some mathematics they lacked a concept of *proof*.

It seems that philosophy and science have flourished best in small mercantile states that have traded and sometimes fought with one another. Mathematics, astronomy and natural philosophy prospered on the shores and islands of the Aegean despite shifting alliances and almost constant warfare between Athens, Sparta, Macedon and Persia.

The Italian Renaissance rose against a background of conflicts between Venice, Genoa, Tuscany and Rome, and the Islamic conquests of Byzantium, Asia Minor and North Africa. The resurgence of art and science was stimulated by contacts with Constantinople, Islamic scholars in the Levant and resulted in neo-Platonism and a revived interest in Greek mathematics, mechanics, astronomy and medicine.

The scientific enlightenment in north-western Europe occurred during a period of vicious religious wars in the Netherlands, German principalities and England. Here renewed interest spread from the universities of north Italy but also from Spain, which had a strong Moorish inheritance. All of these small states whether in the Aegean,

Italy or around the North Sea depended on a skilled yeomanry; artisans and sailors, navigators and traders who had a sense of their own worth and knowledge and were prepared to fight when needed.

Perhaps a measured amount of conflict inspires a spirit of inquiry, exploration and innovation favorable to scientific endeavor. Yet scientific activity requires quiet time for thinking and experiment, whereas successful arms and commerce are noisy, aggressive and impatient. Perhaps the pairing of quieter university cities with ports - Padua with Venice; Florence with Pisa; Leiden with Amsterdam; Cambridge and Oxford with London - overcomes this paradox.

* * *

Prior to the development of experimental science, physics in its **first phase** meant natural philosophy in general, especially the dialectic Aristotelian system of knowledge that may now be regarded as proto-scientific. It was teleological in character and did not have sufficient explanatory power to understand biological phenomena.

This gave ground in the **second phase** to the systematic and objective physics of Archimedes, Galileo and Newton. The knowledge of solid and fluid mechanics were the initial keys to understanding human motion and the circulation of the blood. The conjectures of Boyle and the chemical physicists helped explain respiration and metabolism; the discoveries of Galvani and the electricians revealed aspects of nerve conduction and muscle contraction; optics, the understanding of vision.

The **third phase** was marked by the rise of the engineer who applied the principles of physics for directly useful purposes; engineers sometimes claim to be *"physicists who are useful."* In this phase, which continues today, physicists, engineers and physicians have collaborated to devise instruments necessary for studying human physiological processes – manometers and thermometers; microscopes and endoscopes; x-ray and ECG machines; medical isotopes and body scanners.

The concepts of physics, which underlie medical technology, have also been used to produce weapons of war. Military and medical developments have often been linked. As surgical instruments and weapons are both made of steel; ultrasound is used to detect either foetal abnormalities or enemy submarines; medical isotopes come from the same reactors as weapons-grade plutonium and computerized tomography of the brain was developed at the same electronics laboratory that researched rocket-guidance systems during the Cold War. Could the same computers have been used? Perhaps computers are the link between *Rocket science* and *Brain surgery*?

LIVING MATTER

PHILOSOPHY AND MEDICINE IN ANCIENT GREECE

Hellenic philosophy is believed to have first arisen at Miletus on the Aegean shores of Asia Minor where Thales (640 – 546BC) and his followers established the Milesian school of philosophers. He was probably a merchant who learned of the astronomical and mathematical discoveries of the Babylonians and Egyptians during his travels. Thales proposed that water was the basic matter of which all things are made. Although similar views are to be found in Egyptian mythology, Thales made hypotheses open to rational debate and may have laid the foundations of scientific theorizing on the nature of matter.

The theory that all matter was composed of four ultimate elements – *fire, air, earth and water,* which variously combined to produce different materials, both animate and inanimate, was proposed by Empedocles (493-433 BC), a Greek physician and philosopher who founded the Sicilian School of Medicine. The elements were associated with the qualities of heat, dryness, cold and damp and this theory accounted for everyday experience quite well: sun, rain and soil for crops; sailing; making pots or smelting ores. *Fire, flood, storm* and *subsidence* are still important items in a building insurance.

An alternative hypothesis was presented by the philosopher Leucippus (c.440 BC) from Miletus and developed by his pupil Democritus (460 – 370 BC) of Thessaly on the western shores of the Aegean. Democritus proposed that all matter is composed of an infinite number of minute yet indivisible particles, whose different characteristics and combinations account for the varied properties and qualities of everything in the world, alive or not. The minute particles were indivisible atoms (*atomoi*) in a constant state of motion, colliding and rebounding in the *void*; atomic motion allowed change of material form. *Atoms* and *void* were considered the ultimate realities.

The physician Hippocrates the Great (460-357 BC), so called to distinguish him from three other physicians also called Hippocrates, was born on the island of Cos. Legend suggests that he was probably the son of the physician Heraclides who taught him medicine and a pupil of Democritus who taught him philosophy, ethics and mathematics. Plato's dialogue *Protagorus* celebrates Hippocrates and mentions that he charged his medical students as well as his patients. He travelled widely and is supposed to have performed remarkable cures in Thessaly and Argos. In Athens he was credited with controlling a plague, possibly the typhoid epidemics of 430-426 BC during the siege by Sparta, which killed Pericles and a third of the city's population.

The *Hippocratic Corpus*, a collection of medical treatises probably by many other authors as well as Hippocrates, was collated in the Library at Alexandria, a century or more after his death. The *Corpus* indicated a rational and observational practice of medicine relatively devoid of superstitious elements and based on experience. The authenticity of Hippocrates' contributions in the *Corpus* is uncertain but it is likely that the treatises on *prognosis, epidemics, fractures* and *dislocations, instruments of reduction* and the *Oath* were written by Hippocrates himself.

The *Corpus* contained an anthology of over 400 medical truths or aphorisms concerning the timing of treatment in a variety of acute and chronic disorders. It covered all aspects of medicine, including the management of head injuries, ulcers, fistulae, haemorrhoids and midwifery and was clearly not the work of one physician but more likely part of the library of an early Greek medical school.

Knowledge of anatomy and function was derived indirectly from observation of injuries. The Hippocratic treatise on head injuries gave detailed descriptions of the treatment of depressed cranial fractures and infected head wounds. *"For the most part convulsions seize the other side of the body than the head wound and some become apoplectic."*

The *Corpus* centred its explanations of health and illness on the *theory of humours,* body fluids whose equilibrium was necessary for health. *The four humours – blood, yellow bile, black bile and phlegm* had matching attributes – heat, dryness, cold and damp which in turn matched the four *'essences'* or *'elements'* of Empedocles, who may also have initiated the *humoral theory* of living matter. Diseases characterised as due to excessive *humours* were described as *plethoric, bilious, melancholic* or *phlegmatic.* Scrutiny of urine, faeces and phlegm of patients was frequently recommended to assess their state of health. The treatise on the nature of man stated that *"each of the elements must return to its original nature when the body dies; the wet to the wet, the dry to the dry, the hot to the hot and the cold to the cold. The constitution of animals is similar and all the rest too."* The *humoral theory* attempted to link knowledge of the body with the elements of nature.

The treatise on the *Sacred Disease* (epilepsy) was an attack on popular superstitions about the disease and an account of its natural history. *"I do not believe that the Sacred Disease is more divine or sacred than any other. On the contrary it has specific characteristics and a definite cause."*

With our modern knowledge of human biology and disease the *Corpus* presents a strange mix of practical medical wisdom, folk medicine and mythology. The important outcome was that the Hippocratic attitude to illness threw off the religious or superstitious aspects of suffering and made a start towards the scientific understanding of disease. It was compassionate and relieved the sick of blame for their disease; illness was not a punishment for past sins or caused by evil spirits.

The Hippocratic Oath was sworn by medical graduates up until the 20th century and established standards of behaviour for all physicians. Its major tenets were –

"I will use my power to help the sick to the best of my ability and judgement; I will abstain from harming or wronging any man by it.

I will not give a fatal draught to anyone if I am asked, nor will I suggest any such thing. Neither will I give a woman means to procure an abortion.

I will not abuse my position to indulge in sexual contacts with the bodies of women or of men, whether they be freemen or slaves.

Whatever I see or hear, professionally or privately, which ought not to be divulged, I will keep secret and tell no one."

Plato (427-347 BC) was the nephew of Critias who led the thirty tyrants who ruled Athens briefly under Spartan suzerainty in 404 BC. Many of the tyrants had been friends of Socrates and this led to accusations against him when democracy was restored, then to his trial and death. These events were recorded by Plato who used the *Conversations of Socrates* to write distinctive philosophical dialogues. Plato was advisor to Dionysius II, the tyrant of Syracuse, and he narrowly escaped with his life by fleeing from Sicily after a disagreement with the tyrant.

The first impression of Plato's view of reality and the material world seems non-scientific. For him only abstract geometrical forms were real; the objects experienced through the senses were in time and liable to change, were at most impressions, unreliable and not real. The only truly knowable forms were pure geometry; the way from the life of the senses and of vague opinion to the highest philosophical knowledge was through mathematics. Over the door of Plato's Academy was written -

"Let no-one ignorant of geometry enter here."

Plato proposed that matter has a potential to be actualized by form. The quality of a form produces change in matter to serve its purpose. For Plato, the form of the perishable human body was the immortal soul of reason. Such faith in mathematics re-emerged when neo-Platonism reached Florence after the visit of the Byzantine emperor in 1439 and subsequently led Galileo and other Italian professors to examine the mathematical character of natural phenomena. The culmination of the Platonic theory of enduring mathematical form over insubstantial matter must be the geometrical, double helix arrangement of genetic deoxyribose nucleic acid (DNA), which is immortal so long as a species survives, yet controls the replication of generations of individuals of the species which must each go through the cycle of birth, reproduction and death. Perhaps geometry is the ultimate reality and the *key to life* and immortality.

Aristotle (384 – 322 BC) from Macedon came to Athens in 367BC and studied with Plato for 20 years. His father was the court physician to King Amyntas II the father of Philip II and grandfather of Alexander the Great. As a Macedonian aristocrat he would have been as welcome in Athens as a Scottish lord in Elizabethan London. After Plato's death Aristotle left Athens to marry an Aegean princess and to study natural history on Lesbos. He then returned to Macedon where he tutored Prince Alexander who is said to have looked up to Aristotle "*like a father.*"

After the battle of Chaeronia in 338BC, at which the Macedonian army crushed the armies of the southern Greek states, democracy was abolished and the Macedonian monarchy ruled Greece. Alexander the Great became king on the death of his father in 337BC. Aristotle then returned to Athens and founded his own school, the Lyceum, two years later. Many of their letters to each other survive and Aristotle is said to have given Alexander his precious copy of the *Iliad* that he carried with him on his journey to India. Alexander's conquests in Egypt, Persia, Mesopotamia, Afghanistan and northern India spread Greek culture far to the East. Greek medicine spread along the Silk Road to China. Yunnani medicine based on remedies used by Alexander's physicians is still practiced in Afghanistan and North India by hakims who claim descent from them.

Aristotle wrote on many subjects – logic, metaphysics, rhetoric, poetry, biology and physics and sought to explain man and the universe in his treatises on natural things – *ta phusika.*

As a disciple of Plato he held that all objects consist of form and matter. Aristotle thought Democritus was mistaken with his atomic theory, because he believed nature abhorred a *void* or vacuum. His own concepts were more teleological and implied that it is in the nature of objects to be as they are. For Aristotle as for Plato the form of the soul was wedded to the material body; the heart was the seat of intelligence and the soul, "*the first to live and the last to die.*"

In astronomy Aristotle adopted the theory of spheres, first proposed by Eudoxus (408 - 355BC) to explain celestial motion. Eudoxus who had also been a pupil of Plato, theorized that all heavenly bodies, including the sun, belonged to a system of homocentric spheres whose centre was the earth.

Ptolemy (90 – 168AD) tried unsuccessfully to modify the theory to give reliable astronomical predictions and it was not until Copernicus (1473 – 1543AD) proposed his heliocentric hypothesis that observations began to match theory.

Aristotle's view of nature, though hard to reconcile with experimental scientific concepts, had immense influence for two millennia following his death. Unlike his teacher Plato, Aristotle was interested in natural phenomena of all kinds. He collected many specimens of flora and fauna, dissected animals and theorized about mechanics, astronomy, animal physiology and anatomy. From his anatomical dissections he concluded that the mammalian heart had only three chambers, that arteries contained *pneuma* (breath) and that only the veins contained blood.

Although Aristotle's curiosity was encyclopaedic his powers of accurate observation were limited, for which he has been much criticized since the Renaissance; but for all his faults, he established the study of natural philosophy. After Alexander's death and the revival of Macedon's enemies, Aristotle was faced with a charge of impiety in Athens and went into voluntary exile shortly before his death.

PHYSICS AND MEDICINE IN ALEXANDRIA

The Macedonian kings, the Ptolemys, established Alexandria by Egypt in 331 BC and by 300 BC it had become a major centre of learning in the Greek world. Philosophers, physicians, artists, astronomers and geographers lived there. Merchants sailed on the Monsoon winds to South India linking trade to South East Asia and China. It was in Alexandria, at about the same time, that the famous library and museum were built and acted as a public university where experimental physics and the study of human anatomy and physiology were founded as sciences.

The mathematician Archimedes (287 – 212BC) was the first to make serious scientific contributions based on direct observation and experiment. Although he was born in Syracuse and was a kinsman of the ruler, King Heiron II, he visited Alexandria and studied with the successors of Euclid and played an important part in the further development of Euclidean mathematics.

Archimedes calculated a rough value for π and applied mathematics to mechanical engines. He proved the principle of the lever and the law of moments [1] and is said to have claimed '*Give me where to stand, a lever long enough and I will move the World*'.

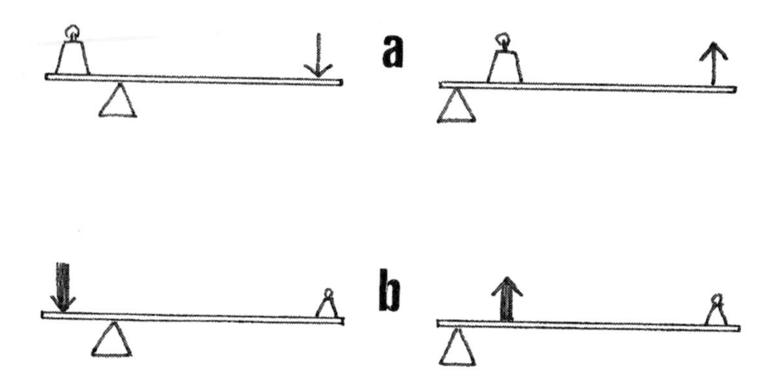

1. Levers. First order levers (a) which use a small force at a large distance from the fulcrum to raise a large weight a small distance or **second order levers (b)** which use a large force close to the fulcrum to move a small weight through a larger distance.

Archimedes recognized that there were two types of levers: **first order levers** use a small force at distance from the fulcrum or pivot to lift a large weight **(a)**, like a crowbar ; **second order levers** use a large force close to the fulcrum to propel a small distal weight at speed **(b)**, like Archimedes' military catapults and ballista.

In his *Hydrostatics*, Archimedes proposed his principle [2]of the equilibrium of floating bodies, having conducted the famous *'Eureka!'* experiment with his own body in a bath. When a body is weighed in air and then in water, the apparent loss of weight is equal to the weight of water displaced.

Archimedes also wrote on inclined planes, centre of gravity and three-dimensional geometry of spheres and cylinders. He established the elements of mechanics on which Stevin, Galileo, Kepler and Torricelli were able to build. He invented a number of contrivances including the water-snail for raising irrigation water, the compound pulley for moving large weights with little force and ballistic instruments to repel the invaders of Syracuse. He was killed by a Roman soldier at the conquest of Syracuse by Marcellus.

The first known anatomist was Herophilus of Chalcedon near the Bosphorus who taught and practiced medicine in Alexandria around 290BC. At that time human dissection was approved in Alexandria, unlike in other Greek cities. The public dissection of human bodies was an innovation of the Ptolemys and remained exclusively the preserve of Alexandrians, possibly because of the familiarity with the Egyptian practice of embalming.

Herophilus was therefore able to make extensive anatomical studies of human cadavers. He made a clear distinction between arteries and veins, and may have recognised that both contain blood and that arteries do not contain air, as had been supposed by Aristotle. His dissections of the cranium identified the lining membranes of the brain, the meninges, and he made the distinction between the cerebral hemispheres and the cerebellum. Unlike Aristotle but like Hippocrates, he regarded the brain as the central organ of the nervous system and seat of intelligence. None of his writings survive and most of what is known derives from Galen, but it appears he also wrote treatises on ophthalmology, the pulse, therapeutics and dietetics.

His younger colleague Erasistratus studied medicine in Athens as a pupil of Metrodorus who was Aristotle's son-in-law. Perhaps for this reason he was strongly influenced by Aristotelian thought, especially the concept of *pneuma*. According to Galen he wrote on anatomy, abdominal pathology, fevers, gout and dropsy. He was a pioneer in the field of pathological anatomy conducting autopsies on men who had just died. Erasistratus was accused by St Augustine of vivisecting criminals.

Herophilus and Erasistratus laid the foundations for the scientific study of anatomy, pathology and physiology. Their careful dissections provided the basis for anatomical investigations by Galen over four centuries later.

Alexandria remained as a major centre of learning even after it was absorbed into the Roman Empire, which adopted much Greek philosophy and most Hippocratic medicine. The Greek physician Galen (130 – 200 AD) and his patron the stoic philosopher and Roman emperor, Marcus Aurelius, both wrote their treatises in Greek. Galen was born in Pergamum in Asia Minor where his father was an architect and mathematician. He had an initial grounding in mathematics and philosophy that made him seek certainty in the study of medicine, above all in knowledge of the human body guided by geometrical principles. He studied medicine in Pergamum, Smyrna, Corinth and then Alexandria where he had the opportunity to study human skeletons, though not cadavers, and dissected apes, cattle, and pigs.

Galen correctly observed that all the veins in the body, except those from the lungs, were connected to the liver, but he viewed blood flow within them as being tidal, ebbing and flowing to and from each organ as it performed its function; the gut to absorb nutrition; the lungs to purify the blood and the brain for animal spirits. Galen did not realize that the blood circulated; that it passed from the heart via the arteries to fill blood vessels supplying all the organs in the body and that blood returned in the veins from those organs back to the heart.

Galen returned to Pergamum as physician to the gladiators having received twelve years of medical education. He moved to Rome in 161 AD where he established his reputation with some dramatic cures. He then became physician to Marcus Aurelius and to the emperors Commodus (180 AD) and Septimius Severus (193 AD). As a result of his own researches and intensive study of the works of Hippocrates, Herophilus and Erasistratus he wrote commentaries and was able to lay down principles in anatomy, physiology and treatment, which dominated for 1400 years; some of his names for bones are still used.

Galen believed that every organ in the body had its own special function and a *'life spirit'* (*pneuma*) pervaded all parts of a living body, expressed as breath, heat and moisture.

In 529 AD the Christian emperor Justinian closed the schools of Athens and suppressed *'heathen learning'* which included the works of Plato, Aristotle, Hippocrates and Galen. From 400 to 1200 AD such few documents as survived in the Christian world were in monastic keeping and regarded with suspicion because of their pagan Greek origins. Hellenic philosophy and science died.

ISLAMIC MEDICINE

Translated into Arabic, Galen's treatises subsequently spread in the Muslim world, which produced outstanding Persian physicians, Al-Razi or Rhazes (860 – 932) and Ibn Sina or Avicenna (989 – 1036) who developed and improved Galen's works.

Al-Razi was an iconoclast who believed that dogma should be questioned and he opposed religious hierarchy. His experience in hospitals in Persia, Baghdad and Basra, especially of fevers, persuaded him that he was a more knowledgeable physician than Galen. He wrote a treatise entitled *Doubts Concerning Galen.*

Ibn Sina wrote *al-Quanun* an encyclopaedia of medicine, which also included contributions from Al-Razi's works. This book was translated into Latin and published in Milan, Padua and Venice in the1480s; it was used as a textbook by the Universities of Montpelier and Louvain until 1650 and was regarded by some physicians as superior to Galen. In addition Ibn Sina produced the *al-Shifa* an encyclopaedia of the sciences based on Aristotle's natural philosophy that covered astronomy, meteorology, natural history, alchemy, mechanics, geometry and algebra.

During the 12th century Christian crusaders became familiar with Islamic medicine in Syria and Palestine. In 1200 AD a medical school was established at Bologna where anatomy was taught for the first time as part of mediaeval medical education by Mondino de'Luzzi (1275 – 1326). His dissections focused on the three chief body cavities: the abdomen, thorax and cranium.

In a restricted form, Aristotle's philosophy and logic were revived in Western Europe by St Thomas Aquinas (1225 – 74 AD) a Dominican friar who stated that reason and religious faith need not conflict but could be in harmony. His attitude to Aristotle's philosophy of nature was that it should be accepted piously and without question. Aquinas' position on faith and reason was subsequently given a favored place in theological teaching and Aristotle became accepted canon for the Church.

Following the introduction of Hellenic ideas by Thomas Aquinas, Roger Bacon (1280 – 1349), a Franciscan, began preaching about the importance of individual experience, both physical and spiritual. Bacon said "*The experimenter should first examine physical things...Without experience nothing can be known sufficiently...Argumentation does not suffice but experience does.*" His disciple William of Occam (1280 – 1349) is known for his logical '*razor*': "*No more things should be presumed to exist than are necessary* "- for proof. Bacon's followers at the University of Oxford discussed questions in physics: the acceleration of free-falling bodies, centre of gravity, levers and optics, so anticipating Galileo by centuries. In 1277 Roger Bacon was condemned by his order and imprisoned.

The Oxford Scholastic Philosophy, which reflected Hellenic culture with its emphasis on free discussion of ultimate questions and its belief in the power of human reason to understand natural phenomena, was denounced as heretical and suppressed.

NOTES

1. ***Archimedes' Law of Levers*** states that ***"a force applied to an arm of a lever, multiplied by its distance from the pivot (fulcrum), equals the weight of the body lifted, multiplied by its distance from the pivot."***
2. ***Archimedes Principle*** states that ***"when a solid is weighed in air and then in a liquid, the apparent loss of weight is equal to the weight of liquid displaced."***

Equal volumes of different substances vary considerably in their ***mass*** and ***weight***. The ***'heaviness'*** of a material is expressed as its ***density***, which is mass per unit volume, measured in ***kilograms per cubic metre*** *or* ***grams per cubic centimetre***.

The ***weight*** of a body is the force of gravity exerted on it at the surface of the earth whereas its ***mass*** is the same everywhere in the universe.

TIME AND MOTION, FLESH AND BLOOD

Physical action and movement are vital attributes for animals. Ingenious machines such as sails and water wheels were able to harness the forces of the wind or of gravity, but until the invention of the steam engine most work was done by skeletal muscle, either human or of domestic animals.

We may differ from other mammals in the relatively large size of our brains and by walking and running only with our hind legs, but otherwise we are very like them. The advantages of the erect posture are that it easier to carry a large head upright and transfer its weight vertically through the cervical spine, at the same time getting a better view. Human hands are freed, so that opposable thumbs give us a unique grasp for using tools, which allow us to use the force of our muscles in more effective ways than with our bare hands. The disadvantage is that unlike other animals, the lower spine, hips, knees and ankles must transmit forces from the whole body to the ground, so making these joints more vulnerable to wear and tear.

The combined scientific advances of physics and physiology over the last two millennia have gradually eroded human 'exceptionalism' based on mythology and taboo, so that it is now accepted that humans are animals; specially gifted but animals never the less, and that the materials of which our living human bodies are composed, comply with the same laws of physics and chemistry that apply to inanimate matter.

Breaking down the myths and taboos, which separated humans from the animal kingdom and from the material world around us started with philosophers and physicians in Ancient Greece. Although Hellenic culture was suppressed for a thousand years, when it was reborn, the teachings of Socrates, Hippocrates, Plato, Aristotle, Archimedes and Galen had a revolutionary effect on theology, philosophy, mathematics, science and medicine.

RENAISSANCE

The renaissance of art in the 15th Century following the revival of Platonism led to an increased interest in the structure of the human body; Leonardo da Vinci (1452 – 1518), Dürer (1471 – 1528), Michelangelo (1475 – 1564) and Raphael (1483 – 1520) were anatomists. Their understanding of the ideal human form with emphasis on proportion and surface anatomy is expressed in their masterpieces of painting and sculpture.

Leonardo was the illegitimate son of a Florentine notary and a peasant girl whose early aptitude for art persuaded his father to apprentice him to Verrocchio in whose workshop he studied painting, sculpture and mechanics. From 1482 until 1499 he was employed by the Duke of Milan during which time he took an interest in mathematics, mechanics and the physics of light. Gravity for Leonardo, as for Archimedes, was the '*force of weight*' straight from a body to the centre of the world.

Leonardo's artistic output in Milan culminated in the fresco *The Last Supper* in 1497. During this period he applied the mechanical principles of the *four powers –movement, force, weight* and *percussion* to human action and sketched men sitting, running and lifting. In the *Anatomical Foleo* he illustrated all the main bones of the body, including the '*lines of force*' of the muscles and the mechanical effects of their leverage in relation to joints.

In his treatise on painting Leonardo said "*Painters study such things as pertain to the true understanding of all the forms of nature's works*" and felt that mathematics held the key to nature and to human physiology. Leonardo studied anatomy for a year at the University of Pavia where he sought geometric '*rules*' for human anatomical proportions. One of his most famous drawings was of *the Vitruvian Man,* which demonstrated that the width of a man's outstretched arms is equal to his height. The work of Leonardo's younger colleague, Michelangelo, was also suffused with the Platonic conviction that outward human appearance embodies inner virtue; that physical splendour reflects fundamental truth.

Leonardo was ahead of his time artistically, in mathematics, science, especially anatomy and physiology. Unfortunately his notes and drawings were in disorder or lost after his death and because he did not publicize his knowledge whilst alive, his impact on the scientific thinking of the time was limited. Leonardo's paintings are a testament to his genius.

VENICE AND PADUA

In the 15th century the Venetian *'Empire of the Sea'* extended to the islands and shores of the eastern Mediterranean, including Athens. It had contacts with Byzantium and the Muslim world and was more independent than other Italian cities. Its government had democratic features similar to ancient Athens, ruled by an elected Senate with the Doge as the first among equals. By necessity it was more innovative and welcoming to outsiders with skills and talents than other cities.

Venice was not a *'daughter of Rome'*. On the contrary in 1509 the Soldier Pope Julius II declared war on Venice and took its *terra firma* provinces of Padua, Vicenza and Verona. Venice recovered these provinces but remained suspicious of Rome. In 1513 France and Venice signed a treaty against their common enemy, the Pope.

A printing press was established in Padua in the 1480s and by 1495 there were 150 presses in Venice, more than in the rest of Italy. Perhaps for these reasons Padua, the university city of the Venetian Republic, became so important for early medical and scientific discoveries. The first Latin edition of Galen's works was published in 1490 and led to a medical reawakening in the 16th century when Galen became known as *"the Prince of Physicians."*

AndreasVesalius (1514 – 1564) was an anatomist from the Netherlands who edited the works of Al-Razi and became professor of surgery and anatomy at Padua in 1538. In 1543 he wrote a great anatomical text, *De humani corporis fabrica (On the Fabric of the Human*

Body). The comprehensive study of the structure and mechanical function of muscles throughout the human body was outstanding.

"I am persuaded that the flesh of muscles is the chief agent by which (the nerves, messengers of the animal spirits not wanting) the muscle becomes thicker, shorter and moves the part to which it is attached."

Vesalius established the teaching of anatomy by human dissection and *Fabrica* provided a new factual basis for medical education. The ambition of medical practitioners was to understand how the body was structured and how it functioned so as to treat injuries and disease.

From his direct knowledge of human anatomy and perhaps from reading Al-Razi's *Doubts Concerning Galen,* Vesalius realised that much of Galen's anatomy was comparative and derived from animal and not human dissections. Even Galen's osteology was faulty because he described the human mandible as made of two bones and the sternum comprised of seven, as in apes. Galen had described nerves as being hollow so as to carry nervous fluids whereas Vesalius was unable to demonstrate any such passages. In the study of the brain he realised that Galen' description of the arteries supplying the brain was derived from ape anatomy and not human.

Until 1628 Galen's false notion that blood ebbed and flowed from the liver to all the other organs of the body in the veins, was accepted even by Vesalius, whose concept of tidal motion of venous blood was little different from that of Galen. This false concept, though understandable, showed the limits of trying to deduce function from dead anatomy. Nonetheless Vesalius' insistence that dissection was essential for the medical student or physician to understand the human body made Padua the first great centre of comparative and human anatomy

Vesalius also emphasized the importance of physicians examining their patients carefully with their own hands. After the publication of *Fabrica* in 1543 Versalius left Padua and became physician to Emperor Charles V. During his service in the Imperial Army he applied his anatomical knowledge to the practice of surgery and devised new techniques, such as the drainage of lung abscesses. His teaching spread widely and he gained the reputation of being one of the great physicians of his age.

THE NEW PHILOSOPHY

"The new philosophy calls all in doubt,
The element of fire is quite put out;
The sun is lost, and th'earth, and no man's wit
Can well direct him where to look for it."
John Donne – An Anatomy of the World.

There was a great scientific revival that resulted from the commercial and industrial prosperity of the cities of the Netherlands and northern Italy in the 16th Century. Merchants from Florence and Venice traded with those of Bruges, Brussels and Amsterdam and there was also an exchange of scholars such as Erasmus, Caius and Vesalius. The University of Leiden was founded in 1575 by William of Orange, a year after he raised the Spanish siege of the city. It quickly became a centre for European studies and students came from Scotland, England and the rest of Europe.

Simon Stevin (1548 – 1620) was born in Bruges and studied mathematics at Leiden University. He became Quartermaster- General of the Army of the States of the Netherlands, science tutor to the Prince of Orange and established a school of engineers in Leiden. He studied the works of Euclid and Archimedes that had come to the Netherlands via Arabic centres of learning in Spain.

Stevin's chief work in mechanics was devoted to discussions of the theory of the lever, inclined plane and the determination of the centre of gravity of bodies and he was probably the first Renaissance author to continue the work of Archimedes. His book was published in 1586, almost fifty years before Galileo's *Two Sciences*. The book was written in Dutch, which limited its impact, but his ideas may have percolated to Italy.

The friar Paolo Sarpi (1552 – 1623) was born in Venice, entered the Servite Order and studied theology, philosophy and logic. He became head of the Order in Venice and State Theologian in 1606 when he counselled the Senate to defy the bull of excommunication launched against Venice by Pope Paul V. The printers of the city were suspected of covertly publishing and distributing Lutheran texts that offended Rome. Sarpi was excommunicated himself in 1607 but remained advisor to the Senate for another 16 years, surviving serious injuries from an assassination attempt in 1608, which he attributed to the Roman Curia.

Sarpi was the first person in Venice to learn of the telescope invented by the Flemish optician Hans Lippershey, when Lippershey applied in October 1608 to Count Maurice of Nassau for a patent on *"a device to make distant objects appear closer"*. Sarpi had good contacts in the Netherlands and learned of the device within a month. He advised the Venetian Senate not to buy one, but to commission his fellow professor, Galileo, to make a better instrument instead. The Senate agreed and Galileo presented his new instrument to the Doge in 1609 for which he received life tenure as professor of mathematics.

Sarpi was well versed in the Scholastic Philosophy and held William of Occam in high regard. His *Artedi ben pensare* examined the relationship of the senses to cognition that anticipated John Locke's *Essay on Human Understanding.*

Sarpi attended scientific meetings in Padua and exchanged ideas with some of the most celebrated scientists of his time. He kept notebooks on reflection of light, bodies in free-fall, relative weights of floating bodies and the motion of projectiles but how much he discovered himself is unclear. He has been credited with correctly interpreting the function of the venous valves and of the circulation of the blood, for which Harvey later provided experimental proof; it is possible that they met in Padua when Harvey was a student there.

Sarpi knew the physicians; Fabricio, Santorio and William Gilbert. Fabricio de Aquapedente was Harvey's teacher, Galileo's physician and it was he who saved Sarpi's life after the attempted assassination. It is clear that Sarpi was at the centre of the scientific life of Padua and Venice and debates about physics and medicine. He was also in a position to advance their cause with the Venetian Senate.

GALILEO

Vincenzo Galilei (1520 – 1591) was a member of a Florentine patrician family and descendent of the famous physician, Galileo Buonaiuti. The family name was changed to Galilei in recognition of the medical miracles he had performed. Vicenzo Galilei was a mathematician and composer who opposed the conventional theories of music based on the Pythagorian doctrine of pure numbers. He declared that neither the authority of ancient writers or number theories could be valid against the evidence of the musician's ear. With the help of his son Galileo he performed experiments with musical strings tensioned by weights.

In 1589 he stated that *"a given musical interval between similar strings is produced either by different lengths or by tensions inversely as the square of those lengths. The perfect fifth is produced by lengths related as 3:2 or when weights in the ratio 4:9 are hung from strings of equal length."*

Galileo Galilei was born in Pisa in 1564, the last year of the council of Trent which confirmed Aristotelian dogma in the Church. His father wanted him to study medicine and restore the family's fame and fortune. Galileo studied medicine and logic at Pisa and then moved to Florence where he studied mathematics, and was later appointed professor of mathematics at Pisa. Following his father's example he was unwilling to accept statements based on ancient authority alone and required evidence instead.

As a medical student attending Mass in the cathedral at Pisa, Galileo noticed that the frequency of oscillations of the lamp, being swung to spread incense, was constant. He timed the oscillations by his pulse at the wrist. He then tested this observation on other pendulums and found that their constant periodicity was independent of the height (amplitude) of their swing – *isochronism.*

"One must observe that each pendulum has its own time of vibration so definite and determinate that it is not possible to make it move with any other period than that which nature has given it."

He measured the periodicity of pendulums of different lengths and discovered that "*The lengths are to each other as the squares of the times of vibration. If one wishes to make the vibration time of one pendulum twice that of another he must make its suspension four times as long.*" It may have been this discovery which diverted him from medicine towards mathematics.

Before Galileo, the measurement of short intervals of *time* was difficult. Astronomic methods gave accurate measurements of years, months and days; however division of the time of day was arbitrary. Europe followed the Egyptians by dividing the day into 24 *hours;* each *hour* into 60 *minutes;* each *minute* was then subdivided into 60 *second-order-minutes,* now called *seconds.* Sun dials, primitive clocks and hour glasses could give approximations of hours or minutes but seconds were more difficult, which is why Galileo had to use his pulse, even though he knew it was variable. He suggested the use of pendulums as clocks and later in his life designed a simple pendulum chronometer suitable for measuring intervals of *seconds.*

A few years later in Bologna, the Jesuit father Giambattista Riciolli (1598 – 1671) and a team of holy fathers patiently counted the oscillations of four *second-minute* pendulums, measuring around a metre in length, for a period of 24 hours, applying gentle pushes every few minutes to keep them swinging.

The closest result was 86,998 oscillations compared with a true result of 86,400; a heroic effort in the search for precision. Marin Mersenne (1588 – 1648), a Franciscan friar recommended that doctors could use a pendulum to measure variations in the pulse rate and "*how the passions of cholera and other fevers hasten it or retard it.*" Mersenne subsequently used a second-pendulum to measure the speed of sound. His result equivalent to 320 metres per second was less than 10 percent below the currently accepted value.

Galileo's experimental findings on the effect of *gravity* on the velocity of falling bodies, very probably from the Leaning Tower, contradicted Aristotle's theory regarding such bodies. Galileo's criticism of the theory provoked the enmity of local Aristotelians and so he left the city. After a short stay in Bologna, Galileo moved to Padua in 1592, where he was appointed professor of mathematics and taught for the next 18 years.

University scholars were not allowed to marry but at weekends Galileo took the ferry to Venice to be with the beautiful Marina Gamba with whom he had three children. Monteverde was composing in Venice at this time and Galileo, as the son of a composer would have enjoyed the music of the city. He also spent time at the Arsenale talking to ship builders, carpenters, plumbers and armourers. Here he learned about the strength of materials, practical mechanics and hydraulics involved in engineering, ship building and gunnery.

In Padua, Galileo endeavoured to establish the mathematical principles of *motion*, by studying levers, moving bodies, pumps and siphons, and he made massively important contributions in the fields of dynamics, strength of materials and hydraulics. He studied the forces on bodies, their speed and acceleration, either in free-fall, suspended or traversing inclined planes. After many experiments he found that for balls rolling down an incline "*the spaces traversed were to each other as the squares of the times, and this was true for all inclinations of the plane.*"

As with his pendulum experiments, Galileo used his pulse to time the movements; supplemented with musical methods. He realised that gravitational force acting on a body produces *acceleration*[1].

Galileo's final work, *Dialogue on Two New Sciences*, that explained his mechanics and criticized Aristotle, supposedly took place in the Arsenale at Venice.

"*Aristotle supposes bodies of different weights fall with different speeds in the same ratio as their weights. I greatly doubt that Aristotle ever tested by experiment whether it be true that two stones, one weighing ten times the other, if allowed to fall at the same instant from a height of 100 cubits would so differ in speed that when the heavier had reached the ground the other would not have fallen more than ten cubits.*" A cubit is the length of a forearm and measures about one foot.

Galileo was aware that suction pumps were unable to raise water more than 30 cubits but attributed this to a breakdown in the cohesiveness of water and not to the weight of atmospheric air. It was left to his pupil Torricelli (1608 - 16 47) to use mercury in a sealed glass cylinder, which showed that the weight of the air, that is, atmospheric pressure, accounted for the phenomenon. The phenomenon was confirmed in 1648, by Blaise Pascal (1623 – 1662), who took a mercury manometer to the peak of Puy de Dom and noted variations in pressure as he moved up the mountain.

In *acoustics,* Galileo continued the experiments of his father by studying the wonderful resonant phenomena of the strings of the spinet.

"A string which has been struck continues to vibrate and these vibrations cause the surrounding air to vibrate and quiver; then these ripples in the air expand into space and strike the strings of neighbouring instruments. A string tuned in unison with the one plucked is capable of vibrating with the same frequency and finally accumulates a vibratory motion equal to that of the plucked string."

He observed that a resonating glass goblet part immersed by water in a bowl created ripples in the water which doubled in number when the goblet was made to resonate one octave higher.

These experiments revealed a wave theory of sound. In *Two New Sciences* the speaker Salvatore says *"This beautiful experiment enabling us to distinguish individually the waves which are produced by a sonorous body, which spread through the air bringing to the tympanum of the ear a stimulus which the mind translates into sound."*

After 1608 Galileo improved the *refracting telescope,* invented by Lippershey and used it for astronomical investigations, which convinced him of the correctness of the Copernican heliocentric theory. He also discovered Jupiter's moons. Although he is best known for defending the heliocentric theory, much of the work had already been done by Kepler and Tycho Brahe. It was Galileo's demonstration of the phases of Venus that provided further evidence for the theory.

His pro-Copernican treatise published in Rome in 1616 had provoked a warning from the ecclesiastical authorities. Galileo had then promised not to *"hold, teach or defend"* the heliocentric theory; however in 1630 he broke his pledge with a dialogue, *Two World Systems,* supportive of Copernicus. Old and ill, Galileo was summoned by the Inquisition in 1632, imprisoned and made to recant. If you can force an old man with gout to kneel for hours, you don't need a torturer's rack.

Galileo's insistence on the use of direct observations and the use of mathematical reasoning to deduce new results from original observations, established the scientific method in physics, particularly dynamics and astronomy. Maybe his insistence on direct observation owed something to his early training as a physician and also to the musical experiments with his father. The development of instruments for making observations and measurements was also important for his discoveries.

The scientific method of direct observation and deduction soon became important for medical measurement and physiology. An associate of Galileo, the Venetian physician Santorio Santorio (1561 – 1636), became Professor of Theoretical Medicine in Padua in 1611. He was the first to use a thermometer to measure body temperature. Santorio invented a pulsimeter, a hygrometer and constructed a balance in which he was able to live. By measuring the weight of his intake and excreta he observed a loss which he attributed to '*insensible perspirations*' and studied these under varying conditions. He published *Ars de statica medecina* (*Medical Statics*) in 1614.

At about the same time, there was another prominent physician, interested in physics, William Gilbert (1540 – 1603), who made major discoveries regarding *magnetism* and *static electricity.*

The magnetic attraction of iron bodies to lodestone was known to the ancient Chinese and Greeks and the properties of magnets fascinated mediaeval alchemists. Supposed medicinal effects were the cure of gout and headaches. Chinese navigators may have used lodestones but certainly magnetic compasses were used in the Mediterranean after 1200 AD.

Gilbert graduated in medicine at Cambridge, practised as a physician in London, became president of the College of Physicians and physician to Queen Elizabeth. For 20 years he studied all recorded magnetic observations and with experiments of his own he deduced that the Earth was itself a huge magnet. He established the study of terrestrial magnetism. Gilbert demonstrated methods of magnetic induction and experimented with static electricity. He divided solid substances into '*electrics*', such as amber and glass (*electric* is Latin for amber) and '*non-electrics*' such as metals. His review of magnetism in *De Magnete* was published in 1600 and founded the sciences of *magnetism* and *electrostatics.*

In 1621 Francis Bacon (1561-1626) who was educated as a barrister, published *Novum Organum,* which comprised a survey of existing scientific knowledge and a method of obtaining information about Nature by performing crucial physical and biological experiments under controlled conditions. Although Bacon tried to establish guidelines for the scientific method, he performed few experiments of his own. He proposed a method for measuring the speed of sound but did complete the experiment.

After the accession of James I in 1605, Bacon had been appointed Keeper of the Royal Seal and Solicitor General. He became Viscount St Albans in the same year *Novum Organum* was published. Bacon knew Gilbert and was a patient of William Harvey, both of whom were royal physicians during his time at court. Intelligent and charming, Bacon had an unpleasant side and was infamous for the examination under torture of a puritan accused of treason in 1614. Was this a method of obtaining information under controlled conditions? Before his death Bacon was imprisoned in the Tower on charges of receiving bribes for favourable judgements. Alexander Pope wrote of him *"think how Bacon shined; the wisest, brightest, meanest of mankind."*

HARVEY AND THE IATROPHYSICISTS

William Harvey (1578 – 1657) studied medicine in Cambridge and Padua, where he graduated in 1602, was lecturer in anatomy to the College of Physicians of London and physician to James I and Charles I. In Padua he was a pupil of Fabricius de Aquapedente (1537-1619), the professor of medicine, who described the valves of the veins and called them *ostiola* or little doors. Fabricio had not immediately recognised the valves' role in making blood flow in one direction towards the heart, but thought at first that they merely damped the tidal flows of venous blood. His book *On the formation of the Egg and Chick* published in 1621 made *embryology* an independent science.

Harvey observed the hearts of live animals and showed that the heart is a muscle and that the atria and ventricles of the heart contract to propel blood through the heart valves into the arteries, which then dilate to give a pulse.

The idea that the heart is a pump was a new concept, and one that he may have derived indirectly from Galileo or Sarpi when he studied in Padua. Harvey showed that there are no pores in the septa of a normal heart; that the whole of the blood passes from the right ventricle through the lungs to the left atrium and ventricle; that all the blood then passes into the general circulation and returns via the veins to the right atrium.

The discovery of the venous valves indicated to Harvey that blood flowing within the veins was directed towards the heart and flow was not tidal as had been supposed by Galen and Vesalius. He deduced the existence of capillaries connecting the smallest arteries to the smallest veins but had no microscope to see them. Harvey published *Exercitatio anatomica de motu cordis et sanguinis (An Anatomical Exercise concerning the Motion of the Heart and Blood)* in 1628.

Although Harvey had discovered the systemic and pulmonary circulations, he did not understand fully why blood needed to pass through the lungs. He assumed it was to cool or purify the blood by contact with the breath but was not strongly convinced.

In *de mortu cordis et sanguinis* Harvey made mention of heart sounds heard on the surface of the chest. *"It is easy to see when a horse drinks that water is drawn in and passed to the stomach with each gulp, the movement making a sound. So it is with each movement of the heart when a portion of blood is transferred from the veins to the arteries, that a pulse is made which may be heard in the chest."*

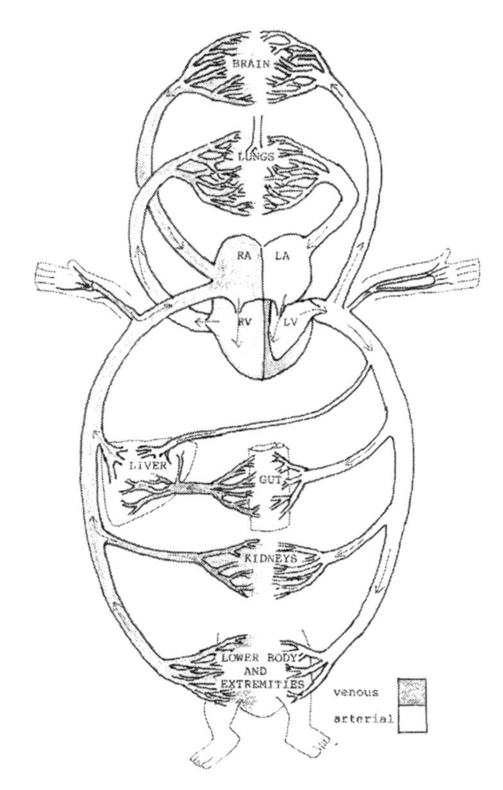

2. The Heart and Circulation of the Blood as demonstrated by **William Harvey. Richard Lower** later discovered that blood changed from venous (blue: shaded) to arterial (red: unshaded) after passing through the lungs. **RA & LA:** right & left atria, **RV and LV:** right and left ventricles.

The execution of his patient, King Charles I, in 1649, must have affected Harvey very greatly. He resigned his medical practice, declined the presidency of the College of Physicians and turned his attention to the embryological development of animals, particularly the detailed study of the development of the hen's egg, following the work of Fabricius. He also described the generation of deer. *Exercitationes de generatione animalium (Exercises Concerning the Generation of Animals)* was published in 1651. By making detailed observations and drawing careful conclusions Harvey established a scientific approach to the function of the living body.

A year after the publication of *de motu cordis* and three years before Galileo broke his promise to the Pope, Rene Descartes (1595 – 1650), after a brief career as a soldier, settled in the Netherlands where he studied for 20 years in the pursuit of philosophical certainty. He resolved to hold nothing as true unless the reasons for certainty were known. This resolve reflected that of Galileo in Padua. Cartesian doubt was to be the instrument of philosophical enquiry and was distinct from scepticism.

Renowned as a mathematician and philosopher, Descartes also had a strong interest in physiology and believed that bodily function could be explained in mechanical terms. During his time in Amsterdam, Descartes dissected animal carcases and is said to have attended the human dissections performed at the Guild of Surgeons by Dr Nicolaas Tulp, famously that of the thief Aris Kindt in January 1632, depicted by Rembrandt in his canvas *The Anatomy Lesson*. Descartes' discourses on vision, *The Dioptrics,* were published in 1637 and *Principia Philosophia* was published in 1644, illustrating his *Principles in Physics and use of certain Terms*. In *Dioptrics* he presented the *'Sine Law of Refraction'*, which had first been expounded in 1621 by a professor of mathematics in Leiden, Willebrorde Snell (1580-1626).

Descartes believed that the soul was seated in the brain and was involved with intellect, imagination and sensation. Sensation, he recognised was mediated by nerves that extend like threads from the brain to all other parts of the body. Stimulation of the nerve endings in the skin, by touch or heat, created disturbances in nerves that were transmitted to the other ends of the nerves, collected together in the brain, around the seat of the soul. Action he thought was caused by a *'spiritual wind'* that passed down nerves to inflate the muscles and so make them contract. Breathing and swallowing, he thought were the necessary mechanical effects of animal spirits discharged to one muscle group or another.

According to Cartesian doctrine, mind or spirit is pure consciousness whereas matter is merely mechanical and can be understood by mechanistic principles alone. Yet both mind and matter are united in humans. The human body he regarded as a machine accidentally united with a rational soul. He considered brute animals to be simple automata. Cartesian dualism posed the challenge of where and how consciousness acted within the body. Descartes was reluctant to characterize humans as robotic and emphasized the importance of *free will* repeatedly. He believed that only human minds were self-aware and they alone had the power of self-determination.

Giovanni Alfonso Borelli (1608 – 79) became a professor of mathematics at Pisa in 1656 and was an astronomer who first suspected that comets travel along parabolic paths. He followed Descartes's mechanical physiology and was a founder with Malpighi of the *Iatrophysical School of Medicine* which sought to explain all bodily functions by physical laws (*iatros* is Greek for physician). He studied muscle function, the mechanics of respiration and the heartbeat. Borelli deserves better recognition for his physiological discoveries but was published posthumously in 1680. In his *De motu animalium (On the Motion of Animals)* which was written before Newton' laws on force, speed, acceleration and work, he described the integrated action of muscles, and the actions of musculo-skeletal levers around joints from studying the way muscle tendons were inserted into bones.

Archimedes had shown that levers could either be either *first order* where a weak force at a long distance could move a heavy weight a short distance, or *second order*. Borelli showed that muscles acting around joints were *second order levers*. The way in which muscles were attached to bone, traded very strong but short contraction of muscles for the much wider but weaker movements of limbs. A man's arm throwing a stone was similar to one of Archimedes' ballista or catapults, but with contracting muscles replacing tensioned ropes. He wrongly assumed that muscles were inflated by gas when they contracted and that the increase in girth reduced their length. Borelli thought this, despite the absence of gas in healthy muscle.

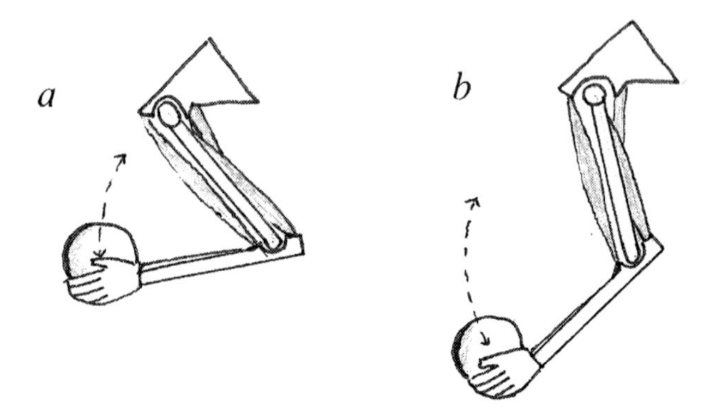

3. Musculo-skeletal levers. Coordinated contraction of biceps and relaxation of triceps muscles in flexion [**a**] and conversely in extension [**b**]. The muscle insertions into the radius and ulna bones of the forearm are close to the elbow joint (**fulcrum**) and trade strong but short changes in muscle length for a large movement of the weight bearing hand. – after Borelli

Borelli was fascinated by Harvey's analysis of cardiac movements in systole and diastole, and the regular involuntary contraction of cardiac muscle, especially the left ventricle. He thought the spiral arrangement of cardiac muscle fibres gave it particular strength and likened the squeezing action in systole to the action of a piston in an engine. He compared the size of the heart muscle to that of the jaw muscles and estimated a force of 150 lbs to propel blood against vascular resistance.

Harvey's work on the circulation was extended by Marcello Malpighi (1628 – 94), who worked in Borelli's anatomy laboratory in Bologna. He became professor of medicine in the university and discovered the capillaries with a compound microscope invented by an optician in the Netherlands. Malpighi studied many organs including the lungs, salivary glands, pancreas, kidneys, the lymphatic system, the spleen, and also the development of the embryo. He came to realize that small blood vessels and networks of capillaries supplied the tissues of all these organs.

THE ROYAL SOCIETY

At the beginning of the English Civil War in 1642, King Charles I made royalist Oxford his capital and his physician, William Harvey was appointed Warden of Merton College during the siege. Large numbers of troops were billeted in the city and there were food shortages and frequent epidemics, particularly of '*Camp Fever*' (typhus). The city surrendered to parliamentary forces in 1646 after the king's defeats at the battles of Naseby and Langport. Immediately Cromwell began to convert royalist Oxford into a puritan city and became Chancellor of the University. College dons were compelled to either change their teaching or be expelled, which started an academic revolution.

A group of thinkers, including physicists and physicians, first met in Oxford during the turbulence of civil war, to discuss experimental philosophy popularised by Francis Bacon in *Novum Organum*. They carried out experiments and dissections in their own houses or college rooms and were known as the '*Invisible College*'. Harvey may have attended some of the early meetings. Prominent members of the group were Robert Boyle, Christopher Wren, Thomas Willis, William Petty and John Wilkins. Robert Hooke and Richard Lower were assistants to Willis and Boyle. The scientific meetings continued, in both London and Oxford, until the Restoration in 1660 when some of the group, with others, formed the Royal Society of London "*for improving natural knowledge by experiments*" which received its royal charter in 1662. The First Charter gave it the right to publish correspondence with foreigners on scientific matters and also the right to demand the corpses of executed criminals for dissection.

Because of his interest in the new experimental philosophy, Robert Boyle (1627 – 91) had moved to Oxford in1654. As a young man he had visited Europe and had studied Galileo's dialogues on *Two World Systems* and *Two New Sciences* in Florence in 1642, the year of the great man's death near the city. During his studies Boyle would have learned about Torricelli's research on pressure and vacuum which was going on whilst he was in Florence. It must have inspired his own interest in air pressure and vacuum.

Boyle opposed Aristotle's notions of '*form and matter*' and was a profound believer in the need to establish an empirically based mechanistic theory of matter and a rational theory of chemistry. Initially he was attracted to medicine and studied medicinal chemistry. Boyle soon became a skilful chemical experimenter and original thinker. With the assistance of Robert Hooke (1635 – 1703), who had constructed an air pump for him, he performed experiments on air, vacuum, combustion and respiration. From these experiments he concluded that air contained a constituent necessary for life and combustion. Boyle observed that sound could not travel in a vacuum and also confirmed Galileo's conjecture that a feather and a lead shot descended with equal speed in a vacuum.

Following Torricelli and Pascal, Boyle made pressure measurements after evacuating different amounts of air from glass chambers. From these experiments he derived a gas law now known as Boyle's Law [2]. He regarded air as having weight and as being composed of elastic corpuscles which made it both compressible and dilatable. Boyle's other major contribution was in the field of chemistry when in 1661 he published *The Sceptical Chemist* which was the starting point of scientific chemistry.

Like other founding members of the Royal Society, Boyle was deeply religious. He studied and translated scripture, and was governor of the Society for the Propagation of the Gospel in New England. Willis, Wren and Hooke were Anglicans who would have become clergymen but for the Civil War which left their families destitute. Wren's father had been Dean of Windsor and Hooke's father a vicar on the Isle of Wight before the Commonwealth. Despite their varied political and religious backgrounds the fellows met weekly to discuss scientific topics, but never religion or politics.

The royalist physician and anatomist Thomas Willis (1621 – 75) studied and practiced medicine at Oxford. William Petty who had studied medicine in Padua during the Civil War taught him anatomy. Willis qualified shortly before Cromwell became university chancellor. Initially he was forced to practise medicine at fairs and in market places but soon built a reputation as an astute physician. He devised new treatments based on iatrochemistry and performed post-mortem dissections on some of his patients to learn the causes of their deaths. At the Restoration he became professor of natural philosophy at Oxford.

Willis was a pioneer in studying the brain and his book *Cerebri Anatome* was published in 1664. In addition to his anatomical studies of human brains he performed experiments on animal brains and nerves. Many of the detailed anatomical drawings of his dissections of the brain, spinal cord and cerebral blood vessels were made by Christopher Wren (1632-1723), the astronomer and architect, who was appointed Savilian professor of astronomy at Oxford in 1660.

Willis described the *corpora striata – the internal capsule and basal ganglia* – as major pathways between the cerebral cortex and brain stem and recognised their importance for sensation and movement. He showed that peripheral sensory nerves from the trunk and extremities connected to the spinal cord and their pathways continued to the brain stem where they crossed to the opposite side, before ascending to the cerebral cortex. Conversely, motor pathways descended from the cortex to the brain stem where they also changed sides and passed via the cord to the peripheral motor nerves, which served voluntary muscles. Cranial nerves arising in the brain stem contained sensory and motor components to serve the head and neck, and these also crossed to the opposite side in the brain stem.

Willis's anatomical demonstration of crossover of nervous pathways explained the Hippocratic observation that injuries of one side of the brain produced fits or apoplexy on the opposite side of the body. Willis also recognised the importance of the optic tract for vision. The major arteries supplying the cerebrum, cerebellum and brain stem, divide and fuse to form a ring within the cranial cavity, immediately below the brain, that is known as the *Circle of Willis,* although he was not the first to describe it. However Willis described it more accurately than previous dissectors and realised its function from the necropsy of a patient who died of abdominal cancer. Willis found that the patient's right carotid artery was blocked and yet he had had no signs of apoplexy in life. Willis discovered that the right vertebral artery had dilated to compensate for the loss of carotid blood flow and that the ring could provide a collateral blood supply.

In addition to the cortico-spinal nervous system which was concerned mainly with conscious sensation and action, Willis and Lower described an additional system of nerves outside the spinal cord pathways, which supplied the heart, blood vessels and other viscera, now known as the *autonomic nervous system.*

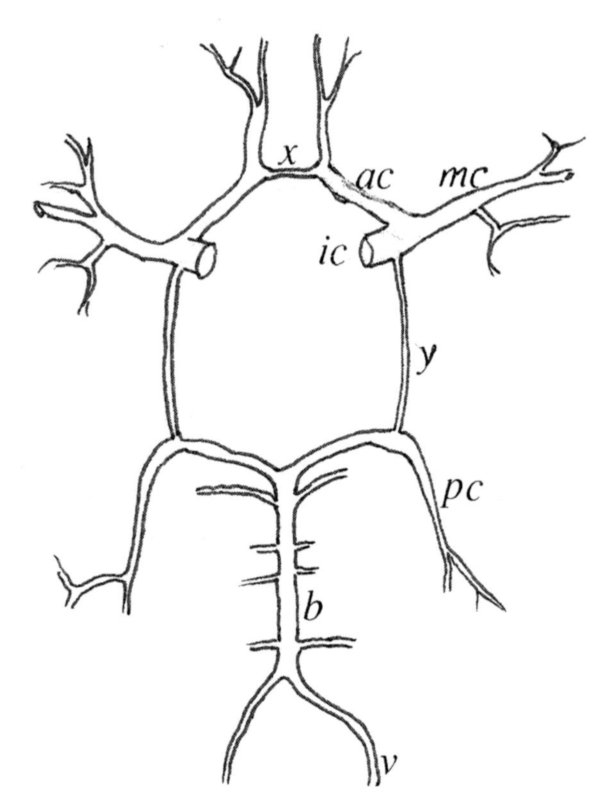

4. The Circle of Willis – the ring of arteries supplying blood to the brain. [**ac**]- anterior cerebral, [**b**]- basilar, [**ic**] - internal carotid, [**mc**]- middle cerebral, [**pc**]- posterior cerebral, [**v**]- vertebral, [**x**]- anterior communicating' [**y**]- posterior communicating.

Willis's knowledge of brain structure allowed him to respond to Descartes's challenge. He speculated that there was a *'corporeal soul'*, common to *'brutes'* and humans, which animated both the blood and nervous juices. He postulated that the nerves were tubes that transmitted nervous fluid from the cerebral cortex, via the brain stem and spinal cord to nerves throughout the body. He regarded the cortex as the seat of memory and imagination and the *'corporeal'* or *'animal soul'* as material, in contrast with the *'rational soul'* that he believed to be immaterial and immortal, consistent with orthodox Anglican theology.

In 1666 Willis moved to London where he had an extensive clinical practice. In 1667 his book *Pathologiae Cerebri et Nervosi Generis Specimen* (*Pathology of the Brain and Specimen of the Nature of Nerves.*) was published.

It included descriptions of hydrocephalus, brain tumours and cerebral vascular disease, derived from necropsies of his patients.

The concept of '*neurologie*' as a discipline within medicine is probably due to Willis. He also proposed the neurological basis for epilepsy and was the first physician to describe myasthenia gravis, which causes severe muscle weakness, as a clinical disease. Although his concept of nerve conduction seems quaint to us now, his discoveries were significant advances.

Perhaps the best English 17th century physiologist after Harvey was Richard Lower (1631 – 91) from Bodmin in Cornwall, who qualified in medicine at Oxford in 1665 and worked with Willis, Boyle and Hooke. Lower became interested in cardiopulmonary function and blood transfusion. Apparently Christopher Wren had previously attempted to inject medicinal liquors into the blood stream of animals using a quill. Lower injected broth into a starving dog's veins with a hollow tube, to keep it alive and told Boyle of his exchange transfusion experiments between dogs "*of equal bigness*".

Lower's first successful demonstration was made in 1665 and in 1667 he joined the Royal Society. Hooke had been interested in respiration since 1664 and thought the efficacy of the lungs depended on their letting air into the body. He blew air through the motionless lungs of a dog to keep it alive. Lower and Hooke attempted to determine the effect on the blood of its passage through the lungs. In the initial experiment they by-passed the lungs and delivered blood from the pulmonary artery to the aorta through an air free channel. Lower saw clearly that the passage of blood through the lungs changed its appearance from blue 'venous' blood to scarlet 'arterial' blood. It appeared 'venous' as it left the right ventricle and was already 'arterial' as it moved from the lungs back to the heart.

Lower had shown that scarlet 'arterial' blood depended on contact with air and not effervescence in the cardiac chambers as had been previously supposed. He also observed the rapid and continuous pumping of the heart with no evidence of effervescence. Lower had proved that the bright red colour of 'arterial' blood "*is due to the penetration of particles of air into the blood*". Willis, who had previously held the '*cardiac effervescence*' hypothesis, immediately applauded Lower and adopted the new concept. Lower cut the cervical arteries of a dog to measure its cardiac output and found that it lost all its blood after only three minutes.

Lower's treatise *Tractus de corde*, published in1669, was quickly recognised as a major new work on physiology. Hooke seems to have dropped out of the research after the beginning, perhaps because he was unable to tolerate the cruelty of the experiments. It is hard to avoid the suspicion that Lower had an underlying compulsive motivation for dissecting the spinal cord or experimenting with blood and exsanguination. This was probably shared by Hooke, even by Christopher Wren. All three were pupils at Westminster School on Tuesday, 30th January 1649, and were locked in that afternoon, so they could not attend perhaps one of the most significant events in English history; the beheading of King Charles I, which took place in nearby Whitehall.

After the execution the Army refused burial of either the body or the head in Westminster Abbey so as to avoid riots. Both were embalmed and buried secretly; however in 1813 they were rediscovered in the crypt of St. George's Chapel in Windsor. Anatomical examination revealed that the axe had struck cleanly through the king's fourth cervical vertebra.

John Locke (1632 – 1704) came to Oxford in 1652 to study theology and became a college tutor in 1660. Locke reacted against Aristotelian scholasticism in support of free enquiry, and preferred facts to words. He also reacted against the intolerance of his fellow puritans at Christ Church. He was affected by the free experimental enquiry that was spreading in Oxford, though not into college lectures. Locke was a student of Thomas Willis, made notes of all his medical lectures and was strongly influenced by him. Locke made medical experiments at Oxford and then practised as a physician. His first university degree was theology not medicine because he refused to accept the conditions for a medical doctorate in 1656. He was subsequently admitted as Bachelor of Medicine in 1675. As a physician he is best remembered for the successful drainage of a liver abscess. The patient Lord Ashley lived for many years afterwards.

Locke developed chronic consumption and after convalescence he became active in public affairs and discussion of social and philosophical problems, particularly sense perception and the limits of human understanding. He was suspicious of abstract speculation and mystical enthusiasm. Locke believed that the essential facts of Christianity were few and simple, but had been obscured by mystical doctrines, most notably the Doctrine of the Trinity adopted at the Council of Nicea in 325 AD. He advanced the causes of religious tolerance and individual freedom.

After 17 years of study Locke's *Essay Concerning Human Understanding* was published in 1690 and its main contention was that what a person can know, or reasonably believe in, must be based on experience. For a man who claimed to prefer facts to words, the *Essay* was quite long and comprised four books and over 700 tightly printed pages. It contained a great deal of psychological and some theological polemic. *"If anyone say, he knows not what 'tis thinks in him; he means, he knows not what the substance is of that thinking thing: no more say I, knows he what the substance is of that solid thing. Further, if he says, he knows not how he thinks; I answer, neither knows he how the solid parts of body are united or cohere together."* The *Essay* established him as the founder of modern empiricist philosophy.

Thomas Sydenham (1624 – 89) was a captain of horse in Cromwell's army who resigned his commission in 1646 to study medicine at Oxford where he knew Locke and encouraged him in his medical experiments. His studies were interrupted by war and he was severely injured at the Battle of Worcester. In 1655 he moved to London where he established his practice as a physician. He made classic descriptions of gout (from which he suffered), venereal diseases, fevers (especially plague and smallpox) and the movement disorder in children following rheumatic fever, which is still known as Sydenham's Chorea. He pioneered the use of cinchona bark (which contains quinine) for fevers in England and used opium in fluid form for easier administration.

Sydenham was also aware of the risks of iatrogenic (physician induced) disorders –

"I act the part of a good physician and an honest man as often as I refrain entirely from medicines when upon visiting the patient, I find him no worse today than yesterday. If I attempt to cure the patient by a method of which I am uncertain, he will be endangered by the experiment and the disease itself; nor will he easily escape two dangers as one."

Sydenham insisted on observation and physical examination in clinical medicine and based his treatments on rational principles, though some were odd, such as *'accubitus'* – introducing a healthy young person into the bed of a devitalised elderly one. He was known in Europe as *'the English Hippocrates'*. Unlike his colleague Thomas Willis he did not perform necropsies or animal experiments.

The contributions to science of Robert Hooke (1635-1703) are hard to assess because of his many collaborations with other great scientists of his acquaintance and the broad spread of his researches. He was both a scientific theorist and maker of scientific instruments to test his theories. Hooke was also involved in the foundation of the Royal Society and was elected curator of experiments and Gresham professor of geometry in 1665.

Hooke had a strong interest and talent in *optics*. In 1665 he published *Micrographia* that contained beautiful and accurate drawings of animal and vegetable structures as well as a description of the optics of his compound microscope. He was the first to identify *'cells'* in the vegetable tissues he studied because of their compact similarity to monks' cells in a cloister.

Probably in collaboration with Wren he made the first attempt at determining the parallax of a fixed star. Hooke constructed a reflective telescope for his astronomical work based on the 1661 design of John Gregor and known as the Gregorian telescope.

The use of curved mirrors overcame some of the problems of refractive telescopes.

Hooke experimented with *sound* as well as light and was aware of its 3-dimensional wave-like or *undulatory* nature.

He observed that -

"The Sense of Hearing does not altogether so much instruct as to the Nature of things as does the Eye, though there are many Helps that this Sense can afford by a greater Improvement, there may be a Possibility that by Otoacousticons many Sounds may be made sensible...There may be also a Possibility of discovering...the Motions of the Internal Parts of Bodies...the Works perform'd in the several Offices and Shops of a Man's Body, and thereby discover what Instrument or Engine is out of order."

What an *'Otoacousticon'* was is not clear but Hooke did suggest *'Helps'* for the auscultation of heart beats. In this he may have anticipated the clinical stethoscope which was definitely established in medical usage in 1819 by the French physician Rene Theophile Hyacinthe Laennec (1781 – 1826). Whilst examining a young female patient with suspected heart disease, Laennec's desire to apply his ear directly to her chest was rendered inadmissible by the age and sex of the patient. He used a paper cylinder instead and was surprised and pleased to find that he could perceive the action of her heart more clearly than by direct application of his ear. But why a stetho***scope*** should be used for hearing sounds, has puzzled generations of medical students. *Otoacousticon* would be a far better name.

There is no doubt that Hooke spread his great talent too thinly over the sciences, from anatomical dissection to cartography, from architecture, astronomy and acoustics to scientific instrument-making. Although he was a capable mathematician, he acknowledged in himself, an inability to reduce '*notions*' to equations. Hooke's ability was in planting ideas rather than bringing them to fruition. As a consequence no major discovery, apart from Hooke's law of elasticity [3], is attributed to him. He almost made discoveries that are now associated with others; *Boyle's Law, Newton's Law of Gravitation, Huygens'* and *Harrison's chronometers* and theories.

Hooke argued publicly with Newton and Huygens over the primacy of his contribution to their discoveries and lost. In 2006, Hooke's minutes of the Royal Society's meetings were discovered in an old trunk at the back of a cupboard in Hampshire and were acquired by the Society. They are likely to reveal much more about this intriguing man. Although Hooke was quarrelsome, he was a close life-long friend of Christopher Wren, John Locke and Samuel Pepys who read and reread *Micrographia*. Pepys was secretary to the Royal Navy, Member of Parliament, courtier and sharp administrator with many connections. Though not a scientist he attended the lectures and dissections of the Society; he became a Fellow in 1665.

Hooke's compound microscope was superseded by instruments made by Anton van Leeuwenhoek (1632 – 1723). He was a Dutch draper with an interest in optics, who made specially ground single lenses which gave greater magnification than previous instruments. Leeuwenhoek improved and extended the knowledge of the capillary circulation, identified red blood cells, spermatozoa, striated muscle fibres and bacteria. He submitted many papers to the Royal Society.

NEWTON

Isaac Newton (1642 – 1727) was born at Woolsthorpe in Lincolnshire three months after his father, who had been a farmer, died. His mother remarried and Isaac was left with his grandparents who gave him a puritan education. As a child he had a talent for making mechanical toys and it seems likely that his intuitive understanding of force, gravity and acceleration came from watching the village blacksmith at work. He was admitted to Trinity College, Cambridge and studied mathematics and natural philosophy especially the works of Galileo and Descartes, and was attracted by Descartes' mechanical philosophy that made him question the nature of force, matter, celestial bodies, light and colour.

Elected to a mathematical scholarship in 1664 Newton had to return to Woolsthorpe for two years during the time that Cambridge University was closed because of the plague. Descartes had regarded a moving body as having *innate force*. Newton regarded a moving body as *passive* and *subject to external forces* acting on it. From this idea he derived the three laws of motion; the effect of forces acting on objects still known as *Newton's Laws of Motion* [4].

In 1683, Halley the Astronomer Royal, Wren and Hooke had became interested in the motions of celestial objects and Hooke speculated that gravitational forces affecting planets might be subject to an inverse square law. They knew from the commentaries of the German astronomer Johann Kepler (1571 – 1630) that the known planets had elliptical orbits around the sun and that the sun was at one focus of each ellipse, but didn't understand why. Halley took the problem to Isaac Newton, who had been appointed as the Lucasian professor of mathematics at Cambridge in 1669 and who claimed that he had already solved this and other problems of planetary motion.

Newton accepted Descartes' assertion that a body in circular motion is forced to recede from the centre by a centrifugal force. Descartes had regarded this force as *innate*. Newton instead regarded *continuous change in velocity* due to rotation as a form of *acceleration*; a force affecting a passive body circling the centre. He observed that bodies; such as apples, are not propelled into space by the centrifugal force of the earth's rotation (the rotational velocity of London is approximately 1000 kilometres per hour) but are instead attracted to the earth by the counteracting force of gravity, which is greater than the centrifugal force.

Newton came to realize that the force of gravity would account for the rotational orbits of the moon around the earth and of the planets around the sun and had the intuition that a body moving under a constant force, inversely proportional to the square of its distance from the centre, would move in an elliptical path.

It took two years obsessively repeating the mathematical calculations necessary to prove this hypothesis to be true. The theory, using his methods of *Fluxions,* predicted the motions of the sun, moon and planets with a high degree of accuracy. Newton accepted the concept of *ether* and inferred that it pervaded all space and matter. He believed it was the medium that transmitted gravitational forces between bodies that were not otherwise in contact.

From this research Newton derived his ***Universal Law of Gravitation,*** which states that "***bodies attract each other with a force proportional to their mass and inversely proportional to the square of the distance between them.***"

During the plague years of 1665 and 1666, working alone at Woolsthorpe, Newton propounded the *laws of motion,* conceived the notion of *gravitation* extending from the earth to the moon, proposed a *theory of colours,* discovered the *binomial theorem* and invented the methods of *Fluxions,* which foreshadowed Leibnitz's publication of differential and integral calculus in 1684 and 1686. Leibnitz's calculus notation was superior to that of Newton who recorded –

"In November 1665 had the direct method of Fluxions and in May following had entrance into the inverse method of Fluxions." Typically, he had failed to publish his inventions before Leibnitz.

This was a remarkable series of discoveries in such a short period and Newton well deserves to be regarded as a genius. In 1675 Newton met Boyle at the Royal Society and adopted his *corpuscular theory,* which was based on that of Democritus.

On his return to Cambridge, Newton equipped a laboratory and carried out chemical and alchemical experiments using sulphur and a variety of metals including mercury. It was said that Newton's furnace in Trinity College burned by night and by day.

He later presented a confidential lecture to the Royal Society on *"Nature's obvious Laws and Processes in Vegetation"* at which he is quoted as saying- *"All things are corruptible. All things are generable. This is very agreeable to Nature's proceedings to make a circulation of all things."*

Newton had a constitutional dislike of publication. It was only with Edmund Halley's persistent enthusiasm and financial support that Newton finally published *Philosophiae Naturalis Principia Mathematica* in 1687 bearing the imprimatur of Samuel Pepys, who had become the president of the Royal Society. *Principia* comprised a systematic treatise on *mathematics, laws of motion, celestial dynamics* and *gravitation.* It also included studies of *fluid mechanics, propagation of sound* and a *law of cooling for heated bodies.* He therefore addressed several aspects of physics; mechanics, gravity, light, heat and sound. He made no recorded observations on magnetism though he mentioned it in his confidential lectures.

In Newton's studies of the fluid mechanics of air and atmospheric pressure [5] he reasoned that if its density was proportional to the degree of compression then densities would increase geometrically towards the earth's surface.

"The density of our own air, as the compressing force; and therefore the density of the air in the atmosphere of the earth, is as the weight of the whole incumbent air that is as the height of the mercury in the barometer." From this proposition he calculated the height of the atmosphere.

Newton observed that *"In experimental philosophy we are to look on propositions inferred by general induction from phenomena as true, not withstanding any contrary hypothesis, till such time as other phenomena occur by which these may be made more accurate or liable to exception."*

Principia was not, and is not, an easy read. Some Cambridge students said of Newton after its publication - *"There goes a man what have writ a book so difficult, no-one, not even he, can understand it."*

However the propositions in *Principia* survived for two centuries until Albert Einstein introduced *special relativity theory* in 1905.

Stephen Hawking, Newton's successor as Lucasian professor of mathematics at Cambridge University, has stated that -

"In fact, the orbits of the planets predicted by (Einstein's) general relativity are almost exactly the same as those predicted by the Newtonian theory of gravity." Except for Mercury, which is closest to the sun and subject to the strongest gravitational effects. Mercury had been noticed, even before relativity theory, to vary slightly from Newton's laws - *"this served as one of the first confirmations of Einstein's theory."*

At the moment of the publication of *Principia*, Newton represented Cambridge University in Parliament against the illegal attempts of James II, to impose his catholic cronies on the university and confronted the notorious Judge Jeffreys. He then became mentally ill in 1692 and 1693. Mercury poisoning from use of the metal in chemical experiments has been suspected as a possible cause. During his illness he wrote letters to John Locke and Samuel Pepys renouncing his friendship with them.

Newton eventually recovered sufficiently to be appointed Warden of the Royal Mint in 1696, which was a tribute to his metallurgical skills as well as his mathematics. He had the responsibility of introducing a new English coinage because over many years of conflict the previous coinage had been debased.

Newton was elected Member of Parliament for Cambridge University and then President of the Royal Society after Hooke's death in 1703. His first edition of *Opticks* finally appeared in 1704; the draft of the original text had been burned in a fire in his laboratory in 1692.

As a person, Newton was undoubtedly strange and his behaviour exhibited features of Asperger's Syndrome, a form of autism. During his optical experiments he blinded himself for several days by trying to look at the sun directly. Then he pushed a bodkin behind his eye to test the effect of pressure on his vision. He communicated poorly or not at all with other college fellows whilst dining and was idiosyncratic in his behaviour to scientific associates. After a conflict with Hooke on colour theory he rejected the study of optics and refused to correspond on the subject. Instead he studied chemistry, alchemy and theology.

The papers on alchemy and his chemical experiments were locked in a trunk with his theological researches and remained undiscovered until 1936, when they were purchased by the economist, John Maynard Keynes. After reading the papers Keynes called Newton *'the last of the magicians.'* Newton was a follower of Arius, an early Christian bishop who had denied the *Holy Trinity* at the Council of Nicaea in 325 AD; however whilst he was a fellow of the *College of the Holy Trinity* in Cambridge he kept these puritan ideas secret, except from his good friend John Locke who shared them. As Warden and then Master of the Royal Mint, which was within the Tower of London, he conducted a merciless campaign against counterfeiting and sent many men to their deaths on the gallows. By this means and with his knowledge of metallurgy and mathematics, he established a sound coinage for England, which Keynes would surely have approved. The reliable currency established by Newton made the City of London a good place to do business. Newton died rich but intestate in 1727; as an Arian he refused the last sacrament of the church. After lying in state he was buried in Westminster Abbey.

The two most influential English thinkers of their age, Locke the physician and philosopher and Newton the mathematician and physicist, communicated frequently on theological and philosophical matters. Both proclaimed the empirical approach to knowledge.

LEIDEN

During the 17th century significant medical discoveries were made at Leiden. Franciscus Sylvius (1614 - 1672) was professor of medicine there in 1658 and dissected the cerebrum; the fissure between the temporal and parietal lobes of the brain is named after him. He was also a chemist and studied digestion and fermentation; he had one of the first scientific chemical laboratories in Europe and was the founder of scientific iatrochemistry. His pupils, Niels Stensen (1638 – 1686) and Jan Schwammerdam (1637 – 1680) both qualified in medicine in 1663. Stensen dissected the tongue, salivary glands, lymphatics, and studied the formation of saliva. The duct of the parotid salivary gland is named after him. He also studied optics and mineralogy. His book *Prodromus* established the principles of geology. His friend Schwammerdam used volumetric methods to study respiration and muscle contraction. He stimulated isolated muscle in a water-filled measuring cylinder and found that it changed very little in volume when it contracted; which disproved Descartes' and Borelli's theory of muscle expansion and shortening caused by '*effervescence*'.

The head of medical school at Leiden in 1709 was one of Newton's greatest admirers, the physician, Herman Boerhaave (1668 – 1738), who based his teaching on clinical history-taking and examination as well as chemical, botanic, microscopic and anatomical instruction. He included aspects of Newton's physics, especially fluid mechanics in his lectures alongside the works of Willis, Lower, Sydenham and Harvey. He performed post-mortem examinations on his own patients, demonstrated to students the relationship between symptoms, clinical signs and pathology. He made significant advances in physiology, clinical medicine and therapeutics.

Boerhaave explained health and sickness in terms of forces, weights and hydrostatic pressures; health was a matter of achieving equilibrium, balancing the pressures of internal fluids. For some this teaching sounded more like that of an engineer or plumber than a physician, in which a person is considered as a hydraulic machine whose pipes are filled with fluids capable of chemical fermentation, whilst the pipes themselves are liable to obstructions or leakages.

At this time, physicians tended to adopt a theory and then fit patients' symptoms to the theory. Boerhaave instead followed the Hippocratic method and it was he who had called Sydenham, *'the English Hippocrates'*. He taught that, as in the physical sciences, medical theories should be based on preceding observations. Boerhaave was the greatest physician and teacher of his time and under his leadership, the medical school at Leiden rivalled Padua. It was a model of scientific medical teaching throughout Europe and the English-speaking world. Boerhaave was elected a Fellow of the Royal Society of London in 1730.

HAEMODYNAMICS

Iatrophysicists applied Newton's mechanics to living systems and indulged in quantitative hydrodynamic arguments based on estimates of blood pressure; size, shape and convolutions of blood vessels; elasticity of vessel walls and resistance to fluid flow, and achieved wildly variable and inaccurate estimates. There was a need for accurate measures of these parameters and Stephen Hales (1677 – 1761) devoted himself to this task.

Hales was a clergyman with no formal medical training but as a student at Cambridge, when Newton represented the university in parliament, he went to lectures with his medical student friends and observed experiments first described by Torricelli, Boyle and Hooke. In 1706, whilst still a student, he performed his first experiment to measure the arterial pressure of a dog. When he became the vicar of Teddington he was fully occupied with parish duties and so did not resume his scientific studies until 1712. He then made direct measurements of the blood pressures in the femoral and carotid arteries, and jugular veins of horses, oxen, sheep, deer and dogs using vertical, tall glass tubes filled with water as manometers to measure the pressures and supposed - *"what is probable,is that the Blood would rise 7 + ½ Feet high in a Tube fixed to the carotide Artery of a Man."* This is equivalent to a systolic pressure of 160 mm Hg and only 40 mm. Hg above what is now regarded as a normal resting blood pressure for humans.

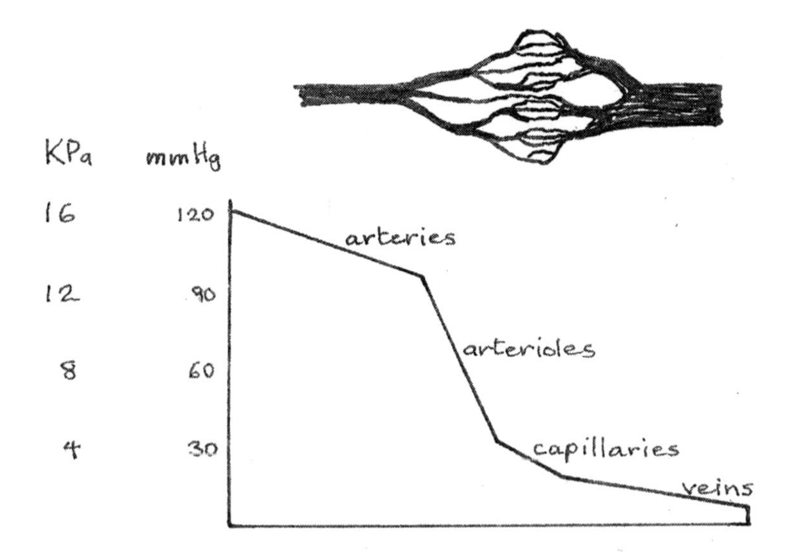

5. The Blood Pressure Gradient through the Human Circulation from Aorta to Vena Cava.

Hales computed the area of the left ventricle to be 15 square inches giving a cardiac force of 50 lbs. He was aware that resistance in blood arterioles and capillaries accounted for the tenfold difference between arterial and venous blood pressures.

Hales had also shown that the resistance to blood flow and the greatest pressure gradient occurs across the small arteries (0.3–10mm. diameter) and arterioles (0.01–0.3mm.diameter). However these small vessels can change their calibre quite significantly and thus have control of tissue blood flow, vascular resistance and systemic blood pressure. Willis in his description of the dense innervation of blood vessels had said the vessels were *"constricted by nervous bridles which moderate the course of the blood according to the impulses of the passions."*

Hales wrote that *"The resistance to blood flow may vary so widely with varying fluidity of the blood and constriction of vessels, never to be exactly the same any two minutes."* He noted that the pulmonary artery and aorta were similar in diameter and blood flow, but that the pressure in the pulmonary vessel was lower than the aorta and that the muscle of the right ventricle was thinner than the left. The force of the right ventricle could not be measured easily because the animal *"must needs dye"* when the manometer was fixed in the pulmonary artery.

Hales was unable to comprehend how arterial pressure contributed to the vigour of muscular contraction and like Willis, supposed the effect of *'animal spirits'* conducted along the nerves. After 1713 he did not pursue the blood pressure experiments in animals any further *"being discouraged by the disagreeableness of anatomical Dissections."* Instead, he successfully transferred his scientific interest in pressure to plant physiology.

Jean Louis Marie Poiseulle (1799 – 1869) studied physics, mathematics and medicine in Paris and was interested in the flow of blood in narrow vessels. He demonstrated that to maintain the same blood flow through a small vessel as a large one, the pressure must increase very significantly [6]. There are approximately 100,000 kilometres of blood vessels in the adult human body varying in diameter from the aorta (25mm) to capillaries (< 0.01mm).

In 1828 Poiseuille improved Hale's method of measuring arterial blood pressure by using a U-tube manometer filled with mercury instead of water to register pressure and anticoagulant fluid to fill the tubing to the artery. As mercury is 13.6 times heavier than water the manometer column was much lower than in Hale's water manometers.

After Poiseuille it became customary to measure blood pressures in millimetres of mercury (mm.Hg). However this method of measuring blood pressure in humans was invasive, inconvenient and not easily repeatable. An easy, non-invasive method of measuring blood pressure clinically was not devised until fifty years later.

The *sphygmomanometer,* or pulse pressure meter, was invented by Samuel Siegfried Karl Ritter von Basch (1837 – 1905) a Czech-Austrian Jewish physician who was the body physician of the emperor Maximillian of Mexico; until Maximillian was executed by a firing squad in 1867. Basch's method used an inflatable pressure cuff to compress the tissues of the upper arm sufficiently to compress and occlude the brachial artery; the main artery in the upper arm. The pressure in the cuff was monitored by a mercury manometer. When pressure in the cuff was reduced to just below systolic arterial pressure, blood would pass into the radial artery at the wrist where a pulse became palpable. The method was improved in 1895 by the Italian paediatrician Scipione Riva-Rocci (1863 – 1937) who used auscultation with the stethoscope over the brachial artery, to identify sound changes associated with systolic and diastolic pressures. This is basically the method which is still used.

COMMENT

After Newton and Boerhaave it became less easy for individuals to pursue multiple disciplines in depth because the knowledge and techniques of physics, biology and medicine had increased and diversified so much. Francis Bacon could claim in 1592 that he would take all knowledge to be his province but by 1700 Newton and his followers were in the ascendancy and there was a strong emphasis on a mathematical view of science. Even John Locke admitted that he couldn't understand Newton's mathematics. The example of Hooke, who spread his great talents too widely and whose mathematics was insufficient to prove his theories, indicated that in future, scientists and physicians would need to focus their research.

The new scientific discoveries and instruments had provided practical benefits such as better time keeping, improved astronomical prediction and navigational aids. Recently invented chronometers and clockwork mechanisms were the wonders of the age and subject to the same sort of adulation we would nowadays give to computers.

These advances derived from the laws of physics which describe *force, momentum, speed* and *acceleration, isochronicity of oscillating systems, fluid mechanics* and *pressure*. These laws had also explained some of the physiological mechanisms of the human body, especially locomotion, acoustics and cardiovascular function.

Two millennia following the foundation of the library and museum in Alexandria, physicians and physicists had gradually developed a good understanding of the mechanical functions of the musculoskeletal and cardiovascular systems. The anatomists, physiologists and iatrophysicists in Italy, France, Netherlands and England had made spectacular advances using quite basic instruments.

By comparison, the understanding of respiration, digestion and metabolism, which are essentially chemical processes, was woeful because as Boyle had discovered, chemistry was still mired in the mysticism of alchemy. Likewise the understanding of nerve conduction and excitation was limited.

A better understanding of the workings of the human body had given no immediate or dramatic improvements in treatments of disease. Most patients were still dependent on herbalists, barber-surgeons or bonesetters for relief of their symptoms or injuries and were at the mercy of fraudulent quacks.

However the bedside clinical methods and theories pioneered by Willis, Locke, Sydenham and Boerhaave were beginning to have a beneficial effect on the care of their own patients and those of their students.

A longer-term, unintended effect of the New Philosophy was to introduce a mechanistic paradigm in place of Aristotle's dogma and Saint Thomas Aquinas' obedience to the divine. Cartesian physiology, which emphasized animals as automata subject to laws of *iatrophysics* was mechanistic and deterministic. *Newton's Laws* reinforced the mechanical philosophy.

If the laws of nature had been defined then the future of the universe was already determined. However the willingness of experimental philosophers to admit that they did not understand particular phenomena was refreshing and was the impetus to design further experiments. There was a scientific optimism that further research would ultimately find answers.

The mechanical philosophy underpinned nearly all scientific and engineering thinking until the twentieth century, when "*other phenomena*" made Newton's propositions of absolute space and absolute time "*liable to exception.*"

NOTES

1. **Galileo** discovered that ***"a motion is uniformly accelerated, when starting from rest, it acquires during equal time intervals, equal increments of speed."***

Acceleration is now defined as the **change in velocity per second** and is measured in **metres per second per second** or **ms^{-2}.**

2. *Boyle's Law* states that '**the volume of a fixed mass of gas is inversely proportional to the pressure'** (provided the temperature remains constant).

3. Experiments with weighted springs revealed ***Hooke's Law of Elasticity*** which states ***"The deformation of a material is proportional to the force applied to it, provided the elastic limit is not reached."***

4. *Newton's Laws of Motion.*
   - I. ***'Every body perseveres in its state of resting, or uniformly moving in a right line, unless it is compelled to change that state by forces impressed upon it.'*** (Every body remains at rest or continues to move in a straight line at a constant speed, if no force acts on it.)
   - II. ***'The change of motion is proportional to the motive force impressed; and is made in the direction of the right line in which that force is impressed.'*** (When a force acts on a body, it will accelerate or decelerate or change its direction of motion. The value of acceleration [or deceleration] is proportional to the force).
   - III. ***'To every action there is always opposed an equal reaction, and directed to contrary directions.'***

The SI unit of force is the **Newton (N)**, which is defined as ***"that force which gives an acceleration of one metre per second per second to a mass of one kilogram."***

5. ***Pressure*** is defined as ***"The force acting perpendicularly per unit area."*** The scientific units of pressure are **Newtons per square metre**

or **Pascals (Pa)** and are quite small; the pressure of one millimetre of mercury (1mm.Hg) is equal to 133 Pa.

6. The French physician and physicist **Jean Louis Marie Poiseuille** showed that -

"The quantity of blood that will flow through a vessel in a given time equals the velocity multiplied by cross sectional area."

$$Q = V \cdot \pi \cdot r^2 \qquad 1$$

Where **Q** equals blood flow, **V** equals velocity and **r** equals radius. He also showed that the mean velocity of laminar blood flow in a vessel is proportional to the square of its radius and the pressure difference across its length - ΔP.

V is proportional to $\Delta P \cdot r^2$ 2
Substituting 2 for V in 1
Q is proportional to $\pi \cdot r^4 \cdot \Delta P$

Poiseuille's principle, derived from his more precise ***Law,*** indicates that ***"the rate of flow of fluid, of standard viscosity, in a tubular vessel is directly proportional to the pressure difference between its ends and proportional to the fourth power of its radius."***

HUMAN ENERGY

Since we first lived in caves humans have known the importance of fresh air, food, water and warmth. These factors and the way they interact are essential to satisfy the energy needs of our bodies. The sun is the source of all the energy available to life on earth and we are dependent on solar energy stored in plant or animal foods to fuel our metabolic requirements. Respiration is the sum of physical and chemical processes by which our tissues use oxygen in energy-producing reactions with nutrients and then excrete waste carbon dioxide and water. Breathing air is a vital part, but only a part, of the whole respiratory process.

AIR

Aristotle and Hippocrates regarded *pneuma* (breath) as the *'life spirit'* and essential for consciousness. Many physicians in the centuries following the Renaissance were naturally concerned with the nature of air, the purpose of breathing and the function of the lungs, because so many of their patients suffered from respiratory illnesses, often attributed to bad airs, malarias and miasmas.

Medical conceptions about air and breathing were confused. The iatrochemists relied on *'chemical principles'* whereas iatrophysicists sought *'mechanical'* explanations. The Swiss physician and chemist, Theophrastus Bombast von Hohenheim, also known as Paracelsus (1493 – 1541), rejected Galen's medicine and introduced a new era of alchemy, diverting it from gold-making to medicine. He developed the *'three principle theory'* of sulphur, mercury and salt; sulphur represented combustion; mercury represented fusibility of metals; salt represented fixity or stability.

The Philosophers' Stone, the medicine of the metals, was the remedy for all ills and the *Elixir of Life*. For Paracelsus breathing was the exhalation of '*nitro-sulphurous airs*' resulting from vital combustion within the body. Many influential physicians such as Thomas Willis in England and Franciscus Sylvius in the Netherlands adopted his principles in their treatments.

The Flemish physician and chemist, Jan Baptista van Helmont (1577 – 1644) was interested in the results of heating chemicals and he may have been the first to use the melting point of ice and the boiling point of water as temperature standards for his thermometer. Van Helmont produced different chemical vapours and condensed solids or fluids from some of them. He likened the different '*airs*' he produced to unformed matter and named them *khaos* after the Greek for unformed or disordered. Apparently his thick Flemish accent made '*chaos*' sound like '*gas*'.

As a physicist Robert Boyle (1627 – 1691) observed that air was elastic when confined in a bladder; that it had weight, and that atmospheric pressure accounted for the difficulty of creating a vacuum. Following his air pump experiments, he regarded air as being composed of permanently elastic particles, which made it '*a thin diaphano-compressible and dilatable Body*', and composed of vapours and exhalations from the earth, water, vegetables and animals.

Boyle and Hooke had shown that '*factitious airs*' could be produced by chemical reactions such as the addition of acids to oyster shells or iron nails. Experiments with lighted candles and live mice in their vacuum chamber showed that both died when air was evacuated and also, after a time, in a fixed volume of air at normal pressure. Hooke made a cabinet large enough to contain himself and experienced shortness of breath, chest and ear pains as air was evacuated. Boyle realized that as well as being elastic, air contained a chemical constituent essential for life and for combustion.

The physician and chemist John Mayow (1640 – 79) initially studied law at Oxford in 1660 and then worked with Boyle, Hooke and his fellow Cornishman, Richard Lower. He was devoted to chemistry, practised medicine in Bath and became a fellow of the Royal Society in 1678. Mayow devised a water trough in which a glass jar was inverted. On a platform inside the jar he ignited antimony and observed that the water level rose due to loss of air and that the antimony ash was heavier than the original metal. He attributed the weight gain to the absorption of *'nitro-aerial particles'* from the air; the residual air could not support the combustion of a candle. His next experiment was with blood in the vacuum chamber. Mayow observed that blue blood did not change colour when vacuum was applied but that arterial blood effervesced and turned from red to blue. He attributed this change to the release of *'nitro-aerial particles'*, which Lower had postulated when he had observed the reverse change in blood colour after it passed through the lungs of a dog. Unfortunately Mayow died shortly after he became a fellow the Royal Society, which was a great loss both to chemistry and human physiology.

The professor of medicine and chemistry at Halle University in Weimar, then court physician to Frederick William I of Prussia was George Stahl (1660 – 1734). He wrote extensively on physiology, pathology and clinical practice but is best known for two concepts; *anima* the *'life force'* which organizes the mechanical functions of the body and *phlogiston* which he regarded as the combustible component of burning material.

Stahl considered flame to be the whirling motion produced by the escape of *phlogiston* and air was necessary for this motion. In the absence of air it could not escape. When it was released into the air it was supposed to pass into plants, be eaten by animals and incorporated in their bodies for vital combustion. Stahl thought soot was a type of pure *phlogiston*. The theory worked well for charcoal, firewood and candles which lost weight when burned, however metals burnt in air produced an ash heavier than the original metal, which indicated that *phlogiston* could have a negative weight. This was confusing because he was almost correct about carbon combustion but deceived by the oxidation of metals.

Although the Reverend Stephen Hales had abandoned animal dissection in 1713 he continued his botanical studies and observed *'air'* in the stems of plants. Air was still considered to be one elementary substance. Gases produced by fermentation, combustion or chemical action were all regarded as *'air'*. Hales repeated Mayow's experiments with a pneumatic trough of his own design, which was shaped like a retort. He was able to measure the loss of air volume when small animals were placed in the trough. Unknown to Hales two effects, the intake of oxygen and the exhalation of carbon dioxide, which dissolved in the water, accounted for the loss. In another experiment he heated red lead ($Pb_3 O_4$) and produced a large quantity of *'air'* and Hales remarked that this could account for the increase in weight when metallic lead was heated to make red lead. Acids acting on salt of tartar (potassium carbonate) also yielded a quantity of *'air'*.

Hales most crucial experiments were made with a rebreathing apparatus made from a bladder and a breathing tube with valves. He found he was only able to breathe untreated air for about a minute because it became *'vitiated'*. If salt of tartar was included in the circuit, especially if causticised (potassium hydroxide) he found that he could breathe for eight minutes. From this he concluded that the soda imbibed *'sulphurous steams'*. As a consequence of his work on different *'airs'* and respiration, he invented ventilators to remove foetid air from prisons, hospitals and slave ships which saved many lives. He was convinced that *'elastic air'*, free of noxious fumes, was essential for healthy respiration.

Hales followed Newton in distinguishing between heat and fire. He believed heat was due to the rapid motion of particles whereas flame is fume heated so much as to emit light. Newton had postulated the exponential dissipation of heat by a hot body, using a primitive thermometer [1].

Daniel Fahrenheit (1686 – 1736) was a German instrument maker working in Amsterdam who made mercury thermometers based on the invention by Roemer in 1708. Fahrenheit provided Boerhaave, the professor of medicine, chemistry, botany and head of the medical school in Leiden with his thermometers. Although physicians in the 17th century knew that the body temperature in health was constant, Boerhaave was able to make more accurate measurements in health and disease, especially fevers, for the first time [2].

The professor of medicine and chemistry in Glasgow, Joseph Black (1728 – 1799), was searching for a lithotriptic agent to dissolve urinary stones, which caused so much painful illness amongst his patients. Black observed that bladder calculi, like chalk or magnesia treated with acid, gave off an *"air that closed a candle"* and that made limewater (calcium hydroxide) turbid (calcium carbonate). He had discovered carbon dioxide, which he called *'fixed air'* and also a method for detecting it. He noted that many chemical reactions produced heat and thought that combustion was caused by the presence of an inflammable principle. Combustible substances, he believed were pervaded with this mysterious substance *"which we are still at loss to explain"*.

Black's engineering colleague at Glasgow University, James Watt (1736 – 1819), made scientific instruments and had been asked to repair a small model of the relatively inefficient Newcomen steam engine installed at the university. As a consequence Watt designed a more efficient and powerful steam engine and obtained a patent for *"a new method of lessening the consumption of steam and fuel in fire (steam) engines."* Watt was astonished by the large amount of cold water needed to condense the steam in the engine cylinder. Black clarified the problem by identifying the *latent heat of vaporization of water* using a Fahrenheit mercury thermometer [3].

Black went on to estimate the specific latent heat for melting ice and discovered the widely differing specific heat capacities of water and mercury. In 1975 he propounded the *'caloric theory'* in which heat or *caloric* was supposed to be an indestructible fluid, which penetrated all bodies and was capable of promoting chemical combinations.

Towards the end of his life Joseph Black wrote " *Chymistry is now studied on solid and rational grounds. Whilst our knowledge is imperfect it is apt to run into error: but Experiment is the thread that will lead us out of the labrynth."*

PRIESTLEY

The development of steam engines by Newcomen and Watt, initially to pump water from mines and to power locomotives, was the start of the industrial revolution in Britain and required larger amounts of coal from ever-deeper mines. Amongst other risks, colliers faced the terrors of *'choke damp'*, *'white damp'* and *'fire damp'*. William Brownrigg, a Cumberland physician who had trained in Leiden, studied the *'airs'* from mines using a pneumatic trough and found that *'choke damp'* extinguished a candle and behaved like Boyle's *'factitious air'*. He found that *'fire damp'* was inflammable. When the gas collected near the roof in deep workings, it exploded when ignited by a miner's candle.

Brownrigg's studies inspired the Yorkshire theologian, preacher, teacher and scientist, Joseph Priestley (1733 – 1804) to make experiments using a mercury-filled pneumatic trough and a burning glass to ignite candles and chemicals held within a jar inverted in the mercury. When he heated mercury ash he saw globules of quicksilver forming and colourless *'air'* rising from it and found that a candle burned in this *'air'* with a remarkably vigorous flame. Priestly repeated the experiment with live mice and found that they survived twice as long in the *'new air'* as in *'common air'*. He had discovered *oxygen*, but in Priestley's time Stahl's *phlogiston* theory was widely accepted so he called it *'dephlogisticated* air'. If he had followed Boyle and Mayow instead of Stahl he might have called it *'nitro aerial spirit'*. Priestley also discovered that the volume of air reduced by one fifth when iron filings and sulphur were ignited in the trough and that the density of the remaining *'air'* (nitrogen) was lighter than *'common air'*.

Priestley's most crucial experiment surprised him most. *"I own I had not that expectation, when I first put a sprig of mint into a glass jar inverted in a vessel of water: but when it had continued growing there for some months, I found that the air would neither extinguish a candle nor was it at all inconvenient to a mouse which I put into it."*

Mint was not a magical ingredient because the experiment worked well with other plants and was most successful with spinach. In 1774 Priestley published the first volume of *Experiments and Observations on Different Kinds of Air* then two more volumes in 1775 and 1777. The president of the Royal Society honoured Priestley in 1772 and said

"From these discoveries we are assured, that no vegetable grows in vain, every plant is serviceable to mankind, making a part of the whole which cleanses and purifies our atmosphere."

Priestley's 'Airs'	**Gases**
alkaline air	ammonia
choke damp, fixed air	carbon monoxide
fire damp	methane
inflammable air(1), white damp	carbon monoxide
inflammable air(2)	hydrogen
phlogisticated air	nitrogen
pure air	oxygen

In 1779 John Ingen-Housz, a Dutch physician in London published his book entitled *'Experiments upon Vegetables, discovering their great Power of purifying the Common Air in the Sun-shine'* which emphasized the necessity of light as well as *'fixed air'* and water, for the most vitally important biochemical process on the planet – the ***photosynthesis*** [4] of carbohydrate, such as glucose, and the release of oxygen into the atmosphere.

LAVOISIER

A new approach to chemistry, developed in France, depended on measuring the weights of components before and after experiments. Antoine-Laurent Lavoisier (1743 – 94) and his colleagues identified *'elements'* which they were unable to reduce any further by chemical means. In 1774 Priestley visited Paris and demonstrated his experiment with mercury ash and *'dephlogisticated air'* to Lavoisier, who had been elected to the French Académie des Sciences at the age of 27. After the meeting, Lavoisier performed a series of experiments and showed that the calcination of all metals was associated with an increase in weight and believed the increase was due to combination with a *'dephlogisticated'* component of air.

He studied the residual *'air'* after calcination and found that it was *'fixed'* but did not cloud limewater like Black's *'fixed air'* and was what is now called nitrogen.

In 1778 Lavoisier announced that *'dephlogisticated air'* was *'oxygen'*. Combustion was *'oxidation'*. The *Méthode de nomenclature chimique* was published by Lavoisier and his colleagues in 1787 and listed the known elements (*substances non décomposées*), which included oxygen, nitrogen, hydrogen, carbon, sulphur, phosphorus and sixteen metals. Although he had abolished *phlogiston* as a substance he had to replace its energetic components and so included *'light'* and *'caloric'* as elements.

Lavoisier did breathing experiments on birds and humans and believed that oxygen was the air that humans and animals inhaled and used up, and *'fixed air'* was exhaled. In 1789, the year the Bastille was stormed and revolution was brewing in the streets of Paris, Lavoisier with his colleague Armand Seguin, showed that human respiration could be the subject of quantitative research and that respiration involved oxygen and not nitrogen, which entered and departed the lungs unchanged. They also showed that oxygen is inhaled and that an equal volume of carbon dioxide is exhaled from the lungs. The reaction in the body was therefore similar to combustion and must be accompanied by the release of *calor* or *'animal heat'*. They found that oxygen consumption increased during digestion and exercise.

Lavoisier observed that oxidation of mercury, iron or lead produced a red calx and suggested that a similar oxidation in the blood turned it from blue to the red colour of arterial blood. Seguin and Lavoisier reported their experiments to the Académie in 1790 and 1791. Like other royal academies, it was abolished during the Terror in 1793. Lavoisier was a minor aristocrat and tax collector, but he was also involved in charitable work to improve conditions in hospitals, prisons and slums. With Benjamin Franklin he was on the committee investigating Mesmer's claims for *'animal magnetism'*. In the final years before his arrest he helped to establish the metric system of weights and measures in France. Surprisingly Napoleon was strongly opposed to the introduction of the metric system and in his memoirs he wrote *"The new system of weights and measures will be a stumbling block and source of difficulties for several generations…It is just tormenting the people with trivia!!!"*

Despite his services to charities for the sick and poor, to science and to France, Lavoisier was executed on the guillotine in the afternoon of 8th May 1794, after trial on the same morning by the Revolutionary Tribunal.

ENERGY AND WORK

Black's *caloric* theory of heat persisted until the mid-nineteenth century when John Prescott Joule (1818 – 89), the son of a Manchester brewer who had studied chemistry with Dalton, proved that electrical energy, heat and work were mutually convertible and formed the basis of the *'dynamic theory'*. He attempted to determine the *'mechanical equivalent of heat'*, that is the amount of heating produced by the expenditure of energy, that is work [5], which he measured in foot-pounds. He designed an apparatus in which falling weights propelled a paddle in an insulated flask of water and measured the rise in temperature for energy expended in rotating the paddle. Joule showed that *mechanical energy* could be converted to *heat energy* [6].

METABOLISM

In the two centuries following Priestley, Ingen-Housz, Lavoisier and Seguin, the chemical understanding of the metabolism of plants and animals increased gradually with improvement in techniques of analytical chemistry. Then in the 20th century with radiochemistry and spectroscopic techniques, there was a more rapid increase with so many researchers making valuable incremental contributions which are beyond the scope of this essay.

However the contribution of Claude Bernard (1813 – 1878) to the study of animal metabolism was outstanding. He was born in the village of St. Julien in Beaujolais and his parents were wine-makers, so from an early age he would have known about the fermentation of sugars. Although Bernard achieved a medical degree in the University of Paris, he never practised as a physician. Instead he devoted himself to physiology, particularly the chemistry of digestion and the transformations of carbohydrates in the animal organism. In 1848 he discovered the presence of sugar in the blood stream of a fasting animal and also the presence of glycogen in the liver. He discovered that glycogen could release glucose and believed that the liver played an important role in maintaining blood sugar levels. Bernard emphasized the importance of homeostatic mechanisms to maintain a constant, physiological environment in the body and established the concept of *milieu interieur* – internal environment.

A simplified overview of the biochemical processes of respiration and carbohydrate metabolism is helpful to understand some of the physics of biological energy. It is easiest to focus on the carbohydrate sugar, **glucose,** which passes readily into the blood stream, is easily taken up by cells and is stored as **glycogen** in the liver and muscles. Glucose has the implicit formula $\mathbf{C_6H_{12}O_6}$, which is equivalent to six molecules of water combined with six atoms of carbon.

Glucose is metabolised to carbon dioxide and water by two processes, firstly by **glycolysis,** which transforms it by a series of reactions into **acetyl coenzyme A,** which is then suitable to join the **carboxylic acid**

cycle. Carboxylic acids are organic acids such as citric acid. The cycle provides a cascade of eight or so chemical steps. These steps involve the incremental **oxidation of carbon to carbon-dioxide** from the **glucose** components and also the release of pairs of **hydrogen** atoms within the cell organelles concerned with energy production: **the mitochondria** [6].

Energy from glucose released within the cell, is captured by **high energy phosphate bonds** in the molecule of **adenosine-triphosphate (ATP),** which is manufactured within the cellular mitochondria [7].

ATP provides the energy for contraction of muscle protein in muscle fibrils, secretions of glands, synthesis of cellular materials of all kinds and transport of substances through the membranes of cells. **ATP is the energy currency of cells.**

Fat and amino acids from protein can be modified to join the carboxylic acid cycle to produce **ATP** for cell energetics. Fat, weight for weight, produces twice as much energy, as carbohydrate. The conversion of **glucose** to **ATP,** through a cascade of chemical reactions looses energy at each step and the ***efficiency of oxidative respiration*** to produce **ATP** is only **44%.** The **remaining 56%** of the energy is released as **heat,** which is why vigorous muscular activity is so warming [8].

Resting muscle cells are able to store significant amounts of ATP as **creatine phosphate,** which is available to fuel contracting muscle with ATP at short notice. Resting muscles contain stored energy, which is quickly available, as both glycogen to providing glucose and creatine phosphate to release ATP. The muscles of trained athletes contain more glycogen, ATP and creatine phosphate than ordinary muscles.

Metabolism is the general term for all the chemical processes taking place in the body and includes the utilization of food for energy, chemical transformations of one material to another, the synthesis of complex compounds for growth and repair and the excretion of waste products. In all these processes heat is produced both in the manufacture of ATP and its utilization so that no more than 25% of all the energy from food is used functionally. This level of efficiency is about that of an internal combustion engine; however respiratory combustion takes place steadily and quietly in the mitochondria of cells and not in the explosive cylinder of an engine.

The basal metabolic rate (**BMR**) is the rate at which energy is consumed in the resting state. For a conscious 70kg human with normal thyroid function, this is about 80 watts and requires the consumption of a quarter of a litre of oxygen per minute. For activity using one litre of oxygen per minute, the power input is 320 watts but produces no more than 75 watts of output; about 200 watts is needed to walk up stairs. Olympic oarsmen can use 4.5 litres of oxygen per minute for ten minutes, equal to an input of 1,500 watts and an output of 370 watts (one unit of horse power is 746 watts).

Dramatic changes between rest and vigorous muscular activity involve significant changes in glucose and oxygen use and the production of large amounts of heat, carbon dioxide and water in contracting muscle fibrils, which contain large numbers of mitochondria.

HEAT

Hermann Ludwig Ferdinand von Helmholtz (1821 – 94) was the son of a Potsdam Gymnasium headmaster who was much influenced by the philosophy of Kant. Helmholtz studied medicine and became a surgeon in the Prussian army. Napoleon wrote from an imperial perspective, although a prisoner on St Helena, that *"an army marches on its stomach."* However he had famously neglected to provide shelter and warm clothing for his troops, which error contributed to his defeat by the Russian winter. Helmholtz as a Prussian army surgeon was probably more interested in foot soldiers' welfare and how much warm clothing, as well as food, a soldier should have to provide and conserve the energy needed to take him and his equipment into battle.

After leaving the army Helmholtz became a physiologist and began to research food, energy, muscle metabolism and thermodynamics. In 1847 Helmholtz published his treatise on *The Conservation of Force,* which was essentially about the physics of muscle. He tried to demonstrate that no energy is lost in muscle movement and that Stahl's *'life forces'* were not necessary for muscle action. Helmholtz then realized that the principle of *Conservation of Energy* applied to all forms of energy. This concept was the foundation of the *First Law of Thermodynamics* [9]. Based on the work of Joule and others he postulated a relationship between mechanics, heat, light, electricity and magnetism by treating them all as manifestations of a single *force* (energy in modern terms). In 1849 Helmholtz became professor of physiology at Kőningsberg, then Bonn and Heidelberg where he did pioneering work on vision and hearing. This brilliant man was then appointed professor of physics in Berlin in 1871; Heinrich Hertz and Max Planck were among his students and became his research associates.

In the same year Helmholtz published his paper on energy conservation, the young professor of physics in Glasgow, William Thomson (1824 – 1907) who later became Baron Kelvin, attended the British Association of Science meeting and heard James Prescott Joule make another of his, so far ineffective attempts, to discredit the *caloric theory* of heat. Thomson was initially sceptical but studied the various thermodynamic theories and corresponded with Helmholtz.

In 1851 in the Journal of Mathematics and Physics, Thomson finally accepted Joule's theory of energy but was concerned about heat loss from thermodynamic systems. Lost heat was not absolutely lost, *"It may be lost to man irrecoverably; but not lost in the material world." "I believe the tendency in the material world is for motion to become diffused, and that as a whole the reverse of concentration is going on,"* which is a definition of ***entropy.*** These ideas led to the *Second Law of Thermodynamic* [10].

Thomson proposed an absolute temperature scale, independent of the physical property of any one substance and postulated there would be a point at which no further *caloric* could be transferred, the point of *absolute zero temperature* [11]. This is now accepted to be equivalent to –273°C; the ice point is 273° on the Kelvin scale.

The concept of '*The Heat Death of the Universe*', as a consequence of entropy is attributed to Helmholtz and perhaps owes something to the fact that he was a physician. Temperature indicates the level of thermal activity in a material. Expansion, which normally occurs during heating, results from the more intense thermal vibrations of atoms and molecules. The realization that all vibrations cease at *zero temperature* raised the possibility of finding relations between molecular size and observable physical quantities. In 1865 the first reliable estimates of molecular size, based on thermal kinetic theory, were made.

Shortly after and independently, Clerk Maxwell (1831 – 1879), professor of natural philosophy in London and Ludwig Boltzmann (1844 – 1906), a young physicist at the University of Vienna, both confirmed that temperatures and heat involved only molecular movement and implied that the concept of *caloric* was redundant. They postulated that molecules within a gas do not have identical energies, even at constant temperature, and that there is a statistical distribution of molecular velocities and energies. Boltzmann defined the mathematical relationship between entropy and probability.

The Maxwell-Boltzmann kinetic theory of gases was a change from the concept of certainty, in which heat was considered to flow only from hot to cold, to one of statistics. Molecules at high temperatures have only a high probability, rather than absolute certainty, of moving towards those at low temperature. This theory was the start of **statistical physics.**

THE INTERNAL ENVIRONMENT

Homeostasis is the maintenance of a stable state within a biological system, by means of internal regulatory processes that counteract external disturbance of the equilibrium.

Active physiological mechanisms keep the internal environment of the human body, and all its cells, remarkably stable and involve transfers of ions, molecules and heat between four different fluid **compartments** – the **intracellular** and **extra-cellular** spaces, the **blood** and the **pulmonary air** spaces.

Homeostasis is not static but is a balanced, dynamic process involving a number of physical and chemical components. Except for blood and blood vessel cells, most tissue cells in the body are not in direct contact with blood but separated from blood capillaries by the extra-cellular space, which contains a weak saline solution. This solution is similar to blood plasma but does not normally contain plasma proteins. Nearly all the cells in the human body are within 30 microns of the nearest blood capillary.

Diffusion is an entropic process in which molecules or ions in high concentration in part of a fluid compartment spread throughout that compartment until the concentration is uniform. **Human body compartments** are separated either by **cell wall membranes, capillary walls** or **respiratory membranes** and diffusion will occur from a compartment with a high concentration to one with a low concentration, provided that the membrane between them is permeable. Transport of glucose from blood plasma via a capillary wall into the extra-cellular fluid and then a cell, such as a muscle fibril, follows a gradient as muscle glucose and glycogen are used up within the cell. Insulin assists glucose uptake by muscle cells. Water also diffuses easily into blood plasma attracted by the osmotic pressure of plasma protein molecules, which are too large to leave the capillaries.

RESPIRATION

Respiration comprises more than the act of breathing, although this is a vital first step. It represents the whole process by which oxygen is taken from the air to all the tissues of the body where glucose and oxygen are metabolized to provide energy, releasing waste-products CO_2 and water, which are then expelled.

It is a homeostatic process that involves-

1. breathing sufficient air to provide adequate **pulmonary ventilation;**
2. **transfer of oxygen and carbon dioxide across the respiratory membrane** between the air-filled pulmonary alveoli and the pulmonary capillaries and
3. **transport of oxygen and carbon dioxide between the lungs and other organs** via the pulmonary and systemic circulations.

The functional units in the lungs concerned with the transfer of gases between air and blood are the terminal air sacs or alveoli which are each less than a millimetre in diameter. The alveoli are surrounded by pulmonary capillaries and **the respiratory membrane** which lies between them is extremely thin, measuring only a few microns in thickness. The total area of the membranes in all the alveoli of both lungs of a healthy human adult is about 70 square metres, which allows ready diffusion of soluble gases, like O_2 and CO_2, between air and blood. Air reaches the alveoli via the bronchial tree, which branches from the trachea (windpipe) into both lungs. The bronchi are not directly involved in gas exchange.

Human metabolism is very dependent on respiratory physics especially on diffusion, driven by differences in gas concentrations and pressures between compartments.

The concentration of a gas diffused in a fluid, either another gas or liquid, can be expressed as its **partial pressure** according to Dalton's Law [12]. Atmospheric air and air in the alveolar spaces of the lungs contain nitrogen, oxygen, carbon dioxide and water vapour. Each gas has a partial pressure proportional to its concentration[13].

A pressure gradient drives the diffusion of oxygen from the alveolar air space into pulmonary capillaries[14]. Transport of oxygen in the blood is more complex than for carbon dioxide because it is only $^{1}/_{20}$th as soluble. Instead of being dissolved, oxygen molecules (O_2) are loosely attached to iron atoms in the pigment **haemoglobin**, in the red blood cells, to form **oxy-haemoglobin**, which is scarlet in colour. When oxygenated blood reaches the tissues, which are using up oxygen and lowering its concentration and partial pressure (**P_{O2}**), **haemoglobin** releases its oxygen for cells to use [14]. Blood returning from the tissues to the lungs in the veins has a low concentration of oxygen and a low **P_{O2}** [14]. **Haemoglobin** becomes **deoxy-haemoglobin** ,which is blue.

Arterial haemoglobin is not an oxide as Lavoisier had postulated but merely an oxygen carrier. The loose binding of oxygen to haemoglobin makes it responsive to changes in the partial pressure of oxygen (**P_{O2}**) in the tissues.

Carbon dioxide is very soluble in blood plasma and unlike oxygen does not require a molecular carrier. The diffusion gradient for CO_2 is in the opposite direction to that for oxygen because the partial pressure of carbon dioxide [**Pco_2**] is highest within cells and surrounding extra-cellular fluid of active tissue. A pressure gradient propels CO_2 from the tissues into the tissue capillaries, then into draining venules and veins[14]. Increased CO_2 in the veins raises their **Pco_2** so when blood enters the pulmonary capillaries, CO_2 diffuses rapidly through the respiratory membrane into the alveolar air space.

Ventilation of pulmonary air spaces with inspired air, with a low **Pco_2** of only 0.3 mm, Hg washes CO_2 out into the atmosphere, so that blood returning to the tissues in the systemic arteries and arterioles has a lower **Pco_2** than the draining veins.

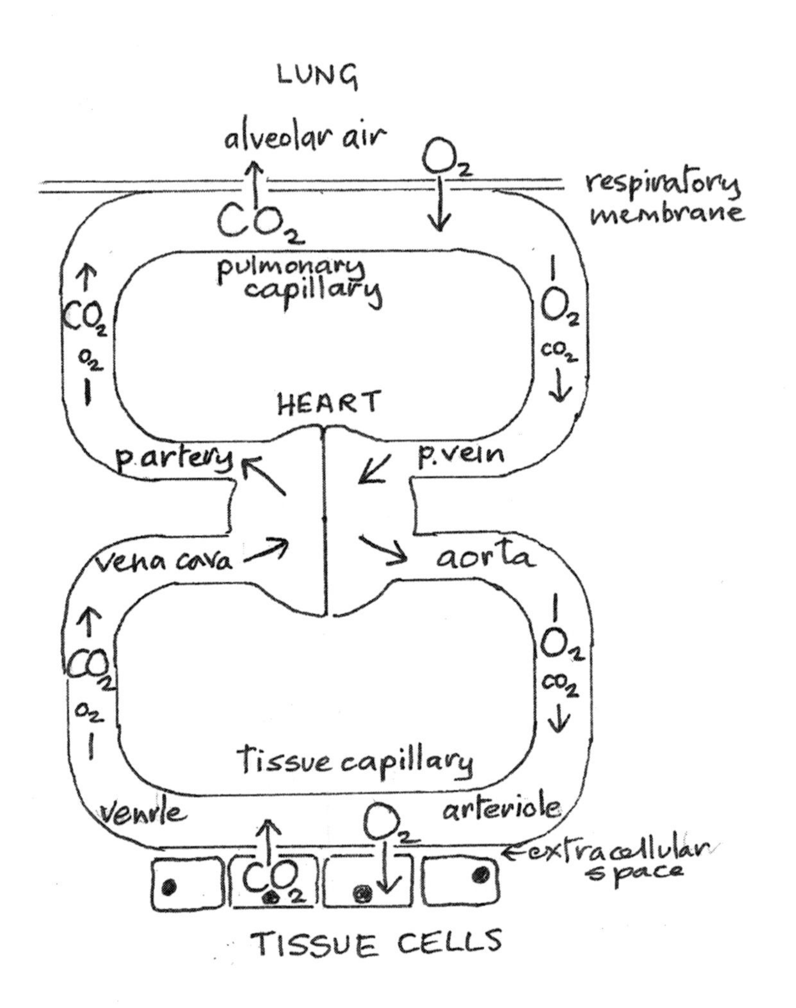

6. Transport of oxygen and carbon dioxide between alveolar air and tissue cells.

Carbon dioxide is soluble and readily transportable in blood. It also has two effects, which have an impact on respiration. Increasing concentrations of CO_2 in tissues promote dilatation of small arterioles and capillaries so that blood flow increases dramatically, bringing more oxygen and glucose to the tissue and taking away surplus CO_2, water and heat. The other effect of an increase in blood $\mathbf{P}_{CO2}$ is to stimulate the respiratory centre in the brain to increase tidal ventilation of the lungs so as to wash-in increased amounts of oxygen and wash-out extra CO_2.

John S Haldane professor of biometry in the University of London, found that a small increase of only 0.2% in the CO_2 of alveolar air (a rise in partial pressure of 1.5 mm Hg), doubled the volume of air breathed per minute [15].

One of the pioneers of respiratory physics and physical chemistry was the physician Christian Bohr (1855 – 1911), professor of physiology at the University of Copenhagen and the father of Niels; creator of the *quantum atom*. He characterized '*dead-space*', which is the volume of air in the tracheo-bronchial tree, which dilutes the exchange of alveolar air with atmospheric air. With good tidal ventilation, dilution is insignificant but alveolar CO_2 concentration ($\mathbf{P}_{CO2}$) will increase to distressing levels if ventilation is diminished, as in asthma, emphysema or paralysis of the respiratory muscles. This is because if ventilation is reduced, only a small amount of fresh air can get through the '*dead-space*' and down to the alveoli.

Christian Bohr recognised the importance of gas diffusion across alveolar and capillary membranes and would have been familiar with the Maxwell-Boltzmann kinetic theory of gases and statistical physics. In 1904 Christian described the phenomenon known as the *Bohr Effect*, whereby the increased $\mathbf{P}_{CO2}$ in the capillaries of metabolically active tissue decreased the binding affinity of oxygen to haemoglobin, so promoting the release of oxygen from oxy-haemoglobin in the vicinity of active cells. Christian Bohr was a strong advocate of accurate quantitative physiological methods and an exemplar of the links between physics, physical chemistry and physiology.

At high altitude; either up a mountain or in un-pressurized aircraft, atmospheric pressure is reduced and so is the partial pressure of oxygen. At 3,000 metres above sea level, atmospheric $\mathbf{Po}_2$ is only 67 mm Hg. At this height haemoglobin carries less oxygen in the blood than normal and this causes breathlessness. The partial pressure of oxygen can be raised by breathing pure oxygen so that breathlessness at altitude can be relieved.

The passage of blood through the circulation brings oxygen and glucose to the tissues and removes carbon dioxide, water and excess heat from them, to maintain homeostasis. Body tissues are vitally dependent on adequate blood flow to maintain tissue respiration. Inadequate blood flow through constricted vessels reduces oxygen delivery and leads to a build up of CO_2. Anaerobic (anoxic) respiration can provide energy for a short time but this leads to the build up of lactic acid, which causes pain in leg muscles and cardiac angina. Blockage of vital arteries leads to death of tissues especially in the brain or cardiac muscle, so as to cause cerebral or myocardial infarction.

BODY TEMPERATURE

Heat generated by metabolism maintains the body temperature at about 37°C but this rises with vigorous activity and falls in extreme cold. However the '*core temperature*' of the major organs such as the brain, heart, liver and kidneys, remains relatively constant in health. Surplus heat is emitted by breathing and through the skin, especially of the head and extremities by increased blood flow from skin arterioles into dermal veins. Heat is lost by radiation, conduction and convection to ambient air, and by sweating.

In cold conditions, the '*core temperature*' is maintained at all costs, but temperatures in the skin and subdermal tissues can fall to ambient levels and then dermal arterioles contract to reduce heat loss. The deep veins of the limbs surround their companion arteries and act like heat exchangers so that cold venous blood returning from the extremities to the '*core*' is warmed. If the core temperature falls significantly, the thermal regulation centre in the hypothalamic region of the brain initiates shivering to promote muscle heat production. The metabolism of the brain is especially vulnerable to hypothermia.

Benjamin Franklin (1706 - 1790), diplomat and prolific inventor, noted the principle of refrigeration by observed that on a very hot day, he stayed cooler in a wet shirt in a breeze than he did in a dry one. In 1757 he was sent to England by the Pennsylvania Assembly to conduct a legal protest on its behalf. On a warm day in Cambridge in 1758, Franklin and fellow scientist John Hadley evaporated ether on the bulb of a mercury thermometer and reduced its temperature from 65°F (18°C) to 7°F (-14°C). He published a letter on '*Cooling by Evaporation*' and wrote *"one may see the possibility of freezing a man to death on a warm summer's day."*

RESUMÉ

Humans have always known the vital importance of breathing, eating, drinking and keeping warm; but how air, food and body warmth maintained life was mysterious until the late 18th century. Before this there had been no shortage of alchemical conjectures based on notions such as *pneuma, life forces,* or *anima,* but most of these were partly magical.

Then scientific experiments began to clarify the nature of animal and plant respiration, the composition of air and chemical elements. After some initial confusion regarding the nature of *phlogiston* and *caloric,* chemistry began to emerge as a respectable science and systematic classification of chemical reactions and elements became accepted. At the same time the development of steam engines revealed the thermodynamic relationship between mechanical work, heat and release of energy by combustion of fuels.

Parallel developments in human physiology began to clarify the nature of respiration, energy metabolism and the importance of a stable internal environment for health.

The molecular constituents of the blood and body fluids vary in concentration within narrow limits maintained by homeostatic control. The temperature and acid / base balance of the tissues are also closely regulated. Systemic physiological mechanisms, involving the cooperative interactions of body tissues, respond sensitively to any condition causing a departure from the normal state and a variety of complex physical and chemical homeostatic mechanisms maintain the relatively unvarying composition of the internal environment. In thermodynamic terms, air and food provide the energy for a healthy living body to combat entropy.

NOTES

1. ***Newton's Law of Cooling*** states that "***The rate of cooling of a heated body is proportional to its temperature excess over its surroundings.***"

2. Fahrenheit adopted Roemer's lowest limit, the freezing point of concentrated salt water solution, as zero (0°F) and the temperature of a healthy human as 96°F, which consequently made the boiling point of water 212°F. Celsius made his scale with a water freezing point of 100° and boiling point of 0°, which was inconvenient, so that the Centigrade scale, which inverted the original scale proposed by Celsius, was adopted instead. However in 1948 the Centigrade scale was officially renamed as the Celsius scale with the freezing point of water at 0°C and the boiling point at 100°C.

3. *Latent heat of vaporization* is the amount of energy required to transform water to steam with no appreciable rise in temperature. Black estimated an equivalent of about 2 kilojoules per gram, which is very close to the true value of 2.3 kJ/gram.

4. The notional chemical equation for ***photosynthesis*** is –

$$\underset{\text{carbon dioxide}}{6CO_2} + \underset{\text{water}}{6H_2O} + \textbf{light energy} \rightarrow \underset{\text{glucose}}{C_6H_{12}O_6} + \underset{\text{oxygen}}{6O_2}$$

5. ***Work*** done is ***the force times the distance moved in the direction of the force.***

The **SI** unit of work and energy is the **joule** (J).

One joule is the work done when a force of one Newton moves through a distance of one metre *and it is also* ***the amount of work done or heat generated when a current of one ampere acts for one second against a resistance of one ohm.***

A machine is any device by means of which a force (effort) applied at a point, can be used to overcome a force (load) at some other point. The ratio of useful work done by a machine to the total work or energy put in, is the **efficiency** of the machine.

Efficiency = (output / input x 100) %.

Power is defined as the work done per second or the amount of energy transferred per second and is measured in **watts. One watt (W) is one joule per second.**

6. The unit of heat in the **cgs system** is the **calorie** (c), which is **the amount of heat required to raise the temperature of one gram of water by one degree centigrade.** The mechanical equivalent of heat is approximately 4.2; in other words 4.2 joules of energy are needed to raise the temperature of a gram of water by one degree. One large calorie or kilocalorie (**C**) is equal to 4.2 kilojoules (**kJ**) and both these units are used by medical dieticians, though **kJ** is the **SI unit.**

7. The chemical formula of **ATP** is complicated but can be simplified to –

ADENOSINE – PO_3 ~ PO_3 ~ PO_4 (~ is a high-energy bond.)

ATP is made by the combination of inorganic phosphate (**i**PO_4) and adenosine diphosphate (**ADP**) in the presence of the enzyme **ATPase** and significant amounts of chemical energy. The amount required to make one mole of ATP from ADP is about 30 kilojoules.

ADP + iPO_4 + ENERGY ATPase ATP

Some ATP is formed from ADP as carbon dioxide is released during the carboxylic acid cycle and more is made with chemical energy released by the oxidation of hydrogen in the mitochondrial membranes, resulting in the formation of water. The mitochondria act like small batteries inside cells, powering the conversion of ADP to ATP.

Glycolysis and the Carboxylic Acid Cycle. Glucose is metabolized by glycolysis to make **acetyl coenzyme A-** which joins the carboxylic acid cycle and is converted to carbon dioxide with the release of hydrogen. Oxidation of hydrogen fuels the conversion of **ADP** to **high energy ATP.**

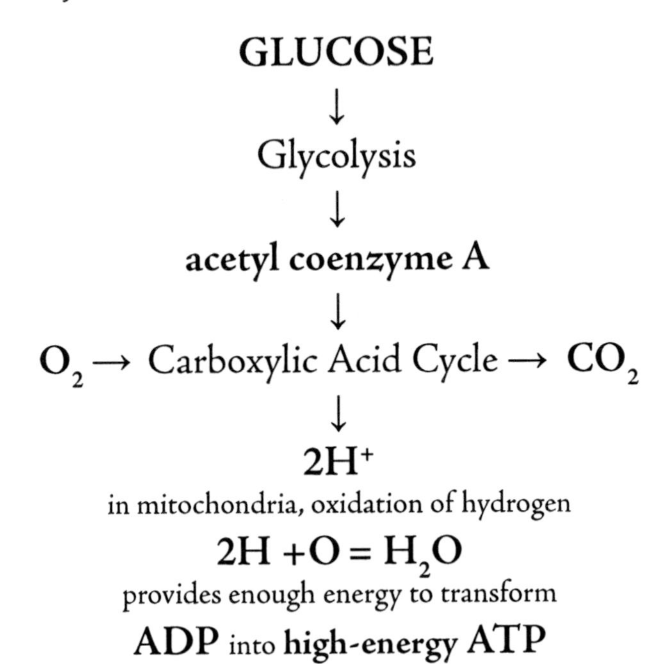

The conversion of ADP to ATP is the rate-limiting step in the respiratory oxidative process and controls the release of energy in relation to a cell's needs. If a cell is not very active it will not use much energy, so only a small amount of ATP needs to be reduced to ADP. Therefore an inactive cell will not have much ADP or iPO_4 available for conversion back to ATP. If a cell suddenly has a large need for energy, say in contracting muscle, it uses up large amounts of ATP and correspondingly large amounts of iPO_4 and ADP will be created. These will activate the enzyme ATPase to manufacture equally large amounts of ATP to replace that which has been used.

8. The oxidation of **one molecule of glucose** during **glycolysis** and the **carboxylic acid cycle** leads to the formation of **38 molecules of ATP** from **ADP.** The molecular weight of glucose is 180 so **a mole of glucose is 180 grams**, equal to about 50 sugar lumps. If these were burnt in a calorimeter they would produce **2,900 kJ of energy**, whereas **38 moles of ATP contain only 1,280kJ**. So the **efficiency of oxidative respiration to produce ATP is only 44%.** The **remaining 56%** of the energy is released as **heat**, which is why vigorous muscular activity is so warming.

9. ***The First Law Of Thermodynamics*** is an extension of the principle of *Conservation of Energy and states* - ***"Energy can neither be created nor destroyed, so that a system can lose or gain energy only to the extent it either passes to or takes from its environment."***

10. ***The Second Law of Thermodynamics***, generally known as the ***Increase of Entropy***, maintains that ***"Physical processes such as the passage of heat from hot to cold bodies, the diffusion of gases and the equalization of concentration gradients in solutions involve changes towards the state of equilibrium."*** Though heat energy can pass from a hot body to a cold one it cannot pass from cold to hot unless additional energy is supplied from outside.

11. ***The third Law of Thermodynamics*** states that ***"At absolute zero temperature the entropies of all crystalline substances are zero."***

12. ***Dalton's Law of Partial Pressures*** *states:*

"The total pressure which a mixture of gases exerts is equal to that which would be exerted by the sum of the individual gases if occupying the same volume." or more simply:

"The partial pressure of each gas in a mixture of gases is proportional to its concentration within the mixture."

13. The partial pressures of gases in air at normal atmospheric pressure of 760mm Hg or 100kPa are -

	Atmospheric Air		**Alveolar Air**	
	mm Hg	**% or kPa**	**mm Hg**	**% or kPa**
N_2	597	78.5	569	74.9
O_2	159	20.8	104	13.6
CO_2	0.3	0.04	40	5.3
H_2O	3.7	0.5	47	6.2

14. In atmospheric air $\mathbf{P_{O2}}$ is 159 mm Hg, which falls to 104 mm Hg in alveolar air and pulmonary venous (oxygenated) blood.

When arterial blood reaches the tissue arterioles the pressure has fallen to 95mm Hg but haemoglobin is still 95% saturated with oxygen. In the tissues oxygen is depleted by respiration, capillary $\mathbf{P_{O2}}$ falls to less than 40mm Hg and at this pressure oxygen is released from haemoglobin to diffuse from the blood into the extra-cellular space, then into the cells.

During vigorous activity the $\mathbf{P_{O2}}$ in returning venous blood is less than 20mm Hg and oxygen saturation of haemoglobin falls to below 30%, in other words it is mainly **deoxy-haemoglobin.** As the blood returns to the pulmonary capillaries the oxygen gradient between alveolar air and the red cells will transform it back into **oxy-haemoglobin** with 97% saturation.

Transport of oxygen and carbon-dioxide between alveolar air and tissue cells follows diffusion pressure gradients.

Partial pressures in mm Hg.

	P_{O2}	P_{CO2}
Alveolar air	104	40
Systemic arterial blood	100	40
Tissue capillaries	95	40
Extra-cellular space	40	45
Inside cells	25	46
Venous blood	20	45

15. The acid/base balance of arterial blood is very slightly alkaline with a pH of 7.4; water is neutral with a pH of 7.0. Carbon dioxide dissolved in plasma makes venous blood slightly more acid than arterial blood with a pH of about 7.35, so wash-out of CO_2 by the lungs restores the acid/base balance to normal levels.

ATOMS TO ATOMS, MYTHS AND MATHS

The atomic hypothesis has been attributed to the philosopher Leucippus (c.440 BC) of Miletus and to have been developed by his pupil Democritus (460 – 370 BC) of Thessaly. It is thought to be have been part of a theory of sense perception. Democritus proposed that all matter is composed of an infinite number of minute yet indivisible particles, whose differing properties accounted for perceived qualities such as colour, texture and fluidity of everything in the world, alive or not.

The minute particles were indivisible atoms (*atomoi*) in a constant state of motion, colliding and rebounding in the *void*; atomic motion allowed change of material form. *Atoms* and *void* were considered the ultimate realities. The conjecture seems to have been more psychological than physical in character and was proposed without any evidence or attempt at proof.

Epicurus (341 – 270 BC) adapted the atomic theory of Democritus to his own concepts of physics and asserted that *atoms* account for the permanence and indestructibility of matter and that *void* allows motion. Epicurus attempted to reassert the atomic theory in opposition to the teleological natural philosophy of Plato and Aristotle. He taught a materialistic theory of nature and the universe, unregulated by divine providence, which considered humans as being like all other living things. Although his name is associated with hedonism, Epicurus and his friends lived modestly in the suburbs of Athens, growing vegetables and enjoying a quiet life. The teachings of Epicurus; materialism and atomic theory, were well regarded by the Roman philosophers Cicero and Lucretius.

Robert Boyle (1627 – 91) studied phosphorus and the properties of acids and alkalis. In 1661 he published *The Sceptical Chemist* which criticized the then current theories of matter comprised of *four elements* (*earth, air, fire and water*) or *three principles* (*salt, sulphur, mercury*). Using his experimental methods he distinguished scientific chemistry from alchemy and defined a chemical element as a substance incapable of being decomposed into simpler constituents. Boyle combined his knowledge of the chemical and mechanical traditions and adopted a *corpuscular* or atomic theory of matter in which all phenomena could be explained by the combination of atoms into mixed bodies, some perfect and some imperfect. Like the alchemists he believed in the possibility of transmutation of base metals into gold but reviled them because of their secrecy and esoteric language.

When Isaac Newton first visited the Royal Society when he was a young professor at Cambridge, Boyle encouraged him to take a scientific interest in chemistry, which resulted in his metallurgical experiments at Trinity College. Boyle had studied Galileo's *Two New Sciences* in Florence in the year of the great man's death and provided an almost direct personal link between these two giants of physics.

Newton adopted Boyle's notion of infinitely small particles of matter. In *Opticks* , published in 1704, he wrote "*It seems probable to me that God in the Beginning form'd Matter in solid, massy, hard, impenetrable, moveable Particles of such Sizes and Figures.... As most conduced to the End for which he form'd them.*"

Throughout the 18th century chemistry began to emerge as a science as researchers in Britain, France, Germany and Scandinavia found pieces of the chemical jigsaw and began to put them together. By mid-century sulphur, phosphorus and carbon had been established as non-metallic elements; then hydrogen by Henry Cavendish in 1766; oxygen by Joseph Priestley in 1774; nitrogen by Lavoisier shortly after. A new approach to chemistry, developed in France, depended on measuring the weights of components before and after experiments. Antoine-Laurent Lavoisier (1743 – 94) and his colleagues identified '*elements*' which they were unable to reduce any further by chemical means.

The *Méthode de nomenclature chimique* was published by Lavoisier and his colleagues in 1787 and listed the known elements (*substances non décomposées*), which included oxygen, nitrogen, hydrogen, carbon, sulphur, phosphorus and sixteen metals. Although he had abolished *phlogiston* as a substance he had to replace its energetic components and so included '*light*' and '*caloric*' as elements. By 1800 over thirty chemical elements had been identified.

Combinations of chemical elements to produce compounds were also revealed; the combustion of hydrogen with oxygen to produce water by Cavendish in 1784; nitrogen with oxygen to produce either nitric or nitrous oxides; hydrogen and nitrogen to give ammonia. In 1794 the French chemist Joseph Louis Proust published his *Law of Definite Proportions* which stated - "*The elemental components of chemical compounds always combine in the same proportion by weight.*"

Shortly before his death in 1799, Joseph Black wrote " *Chymistry is now studied on solid and rational grounds. Whilst our knowledge is imperfect it is apt to run into error: but Experiment is the thread that will lead us out of the labrynth.*"

James Watt's son Gregory joined his father in the engine business but became consumptive and it was considered that living in Cornwall would be good for his health. Gregory Watt stayed with the Davy family in Penzance.

A son of the family, Humphry Davy (1771 – 1829) had been lent one of Lavoisier's books by a shipwrecked French naval surgeon and was fascinated by chemistry. Davy became apprenticed to an apothecary-surgeon in Penzance and was able to perform some of Lavoisier's oxygen experiments. Gregory Watt knew Priestley and had a sound knowledge of the chemistry of that time. Gregory Watt and Humphry Davy became friends; they discussed chemistry and corresponded until Watt's death from consumption in 1804.

In 1789 Davy began his scientific career when he was appointed superintendent to Dr Thomas Beddoe's Pneumatic Institute in Clifton, which specialized in treating consumptive patients and he had the responsibility of preparing and studying new gases ('*factitious airs*') in the cure and prevention of disease. Davy almost died after testing a sample of carbon monoxide and although he was warned not to breathe nitrous oxide, which he prepared by heating ammonium nitrate, he did so, discovered its anaesthetic properties and wrote a book about its possible use in medicine. The Pneumatic Institute failed to find any efficacy for the gas in curing disease but Beddoe's friends, the poets Coleridge and Southey enjoyed its euphoric qualities; it became known as *'laughing gas'* and at the time was regarded as slightly disreputable.

Davy turned his attention to electrochemistry and designed a number of ingenious experiments which led to his discovery of the metal sodium as well as potassium, calcium, chlorine and a few other elements. He is best remembered for designing the ingenious miners' *Safety Lamp* which reduced the dangers of *fire damp* (methane) explosions.

DALTON AND BERZELIUS

After Lavoisier, the two men who did more than any others to establish chemistry as a science, were an English schoolteacher and a Swedish professor of medicine.

Raised in England's Lake District, where the annual rainfall is over three metres, John Dalton (1766 – 1844) became interested in meteorology. He studied the Aurora Borealis, clouds, atmospheric air, other gas mixtures, dew and the precipitation of water vapour. He was the first to establish that rain is caused by a fall in temperature and not a change in atmospheric pressure. Dalton went on to show that gases are soluble in water and that atmospheric air is constant in composition up to a height of 5,000 metres. In his book *Meteorological Observations and Essays* published in 1793, he formulated his *law on the partial pressures of gases* in a volume of mixed gases, such as air.

After his appointment as professor of natural philosophy in Manchester, Dalton extended Proust's research by weighing the different portions of each element in a chemical compound and observed the ratios of weights of different elements. He observed that nitrogen could combine in different ways with oxygen (to produce nitric or nitrous oxide) and could combine with hydrogen to produce ammonia; carbon combined in two different ways with oxygen.

With these experiments, which he pursued to the smallest amount of each element, he confirmed and extended Proust's hypothesis, that chemical compounds were formed from the combination of constant amounts of their elemental constituents.

From these ratios he derived a system of *equivalent weights,* which formed the basis of his *atomic theory* of matter. Dalton converted Greek philosophical *atomism* into scientific theory between 1803 and 1808 when he published *A New System of Chemical Philosophy* .

Dalton's atomic theory proposed that-

- ***Every element consists of very small indivisible particles (atoms).***
- ***Atoms are indestructible.***
- ***The atoms of one element are identical but different from those of other elements in mass.***
- ***Chemical compounds are formed by the combination of atoms of different elements in simple ratios.***

Some of Dalton's measurements were inaccurate; he wrongly assumed that only one atom of hydrogen combined with one of oxygen to form water; some of his concepts were not original. In spite of these limitations, which he subsequently acknowledged, Dalton had established a systematic atomic theory with a strong predictive value. By 1815 it had been revealed that the atomic weights of nearly all the known elements were approximately multiples of that of hydrogen. Dalton's elemental atoms could be visualized as snooker balls of different weights that combined to form molecular compounds held together, as Humphry Davy demonstrated, by electrochemical forces.

Amongst his other discoveries, Dalton found that he and his brother Jonathon were colour blind and John left instructions for his eyes to be dissected, after his death. His DNA was documented using his stored eyes and in 2004 the Dalton Genealogical Society started the Dalton International Gene Pool project to identify his living descendents throughout the world.

As a medical student Jöns Jacob Berzelius (1779 – 1848) worked for a physician at a mineral water spa, analysing the chemical content of the spa water and helping to assess the effects of Voltaic current on patients with different diseases; he was unable to demonstrate any benefits. After qualification he became an assistant to the professor of medicine and pharmacy at the Medical College in Stockholm and in 1807 he was appointed to the post himself. In 1810 the college became the Karolinska Institute. In the same year Berzelius was appointed president of the Swedish Academy of Science and he was able to devote his time to chemistry and to visiting scientists in other countries.

Initially he got on well with Humphry Davy during visits to England but the friendship cooled after Berzelius criticized one of Davy's books. Using electrochemical methods, Berzelius confirmed Dalton's Atomic Theory with more accurate experiments and corrected some of Dalton's inaccuracies. He also showed that the laws of combination applied to organic substances and pioneered the study of carbon chemistry; he coined the terms *protein* (*primitive*) and *catalysis* (*chemical affinity analysis*). Berzelius discovered the elements; cerium, selenium and thorium, whilst students in his laboratory discovered lithium and vanadium. He was an inspired experimental chemist and also a brilliant classifier.

Berzelius's major legacy was the arrangement of all the elements in a series of decreasing **electronegativity. Oxygen**, which combined with everything and was liberated at the positive pole in an electrolytic cell, was obviously **the most electronegative** element, whilst **potassium** was **the most electropositive** known at the time. Paradoxically carbon, which he had studied so carefully, was the most difficult to fit into his classification.

Unfortunately Berzelius did not understand electricity quite as well as Faraday so that physicists did not pay much attention to the theory, but amongst chemists it attracted many followers because it so easily explained the behaviour of inorganic substances.

The other lasting legacy of Berzelius resulted from his study of the chemical compositions of over 2000 chemical compounds and his recognition that a simple, logical system of chemical symbols was required. Prior to this, different notations and different symbols, some derived from alchemy, were used in different countries; Dalton had his own system of circular symbols for elements.

Berzelius designated the first letter of the Latin name of each element, with the second letter if more than one element began with the same letter. At the same time a quantitative concept was included because the Latin notation implied the atomic weight of an element; numbers were included to indicate the proportions of elements in a compound which are now presented as subscripts; of course he was unaware of isotopes. This is the method still used; for example **carbon dioxide** is now written in Berzelian as **CO_2**.

The English chemist, John Alexander Reina Newlands (1837 – 98) made the observation in 1864 that the elements, when arranged in order of their atomic weights, showed a recurrent pattern of chemical properties after each series of eight elements; chemically similar elements recurring in a definite order. He proposed the **Law of Octaves,** which paved the way for the professor of chemistry in St Petersburg, Dimitry Ivanovich Mendeleyev (1834 – 1907), to establish the **Periodic Table of Elements,** which he did in 1869.

ELECTRICAL EXCITEMENT

"I believe that man is the product of the fluids of the air, including the electric; that the brain pumps these fluids and gives life, and that after death they return to the ether."
N Bonaparte.

The physician to Queen Elizabeth I, Dr William Gilbert studied the magnetic properties of lodestones by floating them on corks in a bowl of water and observed that north-seeking poles repelled other north seeking poles but attracted south- seeking ones. He concluded that ***like poles repel and unlike poles attract.*** He also studied what he called *'electric'* effects when amber (*electron*) or glass rubbed with a cloth attracted light objects such as feathers, paper or gold leaf and designed a simple electroscope to demonstrate attraction between charged bodies.

Mechanical devices to increase friction produced stronger charges on sulphur or glass balls, which then gave off *'electric' sparks.* Then in 1706, Francis Hawksbee, Newton's assistant and curator of experiments at the Royal Society, built an *'electric machine'* with a rapidly rotating glass globe which gave off a greenish-blue glow and produced lightning-like sparks when friction was applied to produce *'electric force'*.

ELECTRIC CONDUCTORS

Stephen Gray (1666 – 1736) was a poverty stricken scientist who had worked for John Flamsteed, the Royal Astronomer at Greenwich and had done some good observations of sunspots. He then moved to Newton's Cambridge observatory before it closed because of poor management. Gray was given a pensioned place at Charterhouse, which was an almshouse for destitute gentlemen who had served their country. Here he experimented with static electricity using a glass tube as a friction generator, which had a cork at one end. Gray noticed the cork attracted chaff and bits of paper. He extended the cork with long wooden tapers and then with thread to an ivory ball and observed that '*electric virtue*' would spread over a distance and that the ball attracted light objects just like the glass tube itself.

Flamsteed's relatives and the Reverend Granville Wheler allowed him to extend his experiments through the large rooms and towers of their houses in Kent using thread-wire, which carried the '*electric virtue*' around bends and against gravity for a distance of 800 feet. Gray and Wheler noticed that the thread leaked electric charge and discovered the importance of insulating their thread-wire from earth contact using dry silk. Gray then suspended a Charterhouse charity boy from the ceiling with silk ropes and charged him with the glass tube. Fragments of paper and chaff on a tray below the child were attracted to him. From these experiments came an understanding of the role played by electric conductors and insulators, the release of charge to earth and also that the human body is a spectacular conductor of electric charge. Gray was awarded the Copley medal by the Royal Society in 1731. He was admitted to the society in 1732 but died destitute four years later and was buried in a pauper's grave.

Electric machines became even more powerful and more widely available. Electricity was regarded at this time, like heat or light, as an '*imponderable*', a stream of corpuscles or weightless fluid. The problem was how to store it. In 1746 a professor in Leiden, Pieter van Musschenbroek tried to obtain '*electric fire*' from water in a glass jar, which had been charged with '*electric fluid*' from a machine. A misguided assistant held the jar in one hand, introduced a metal conductor into it with the other and received a big shock.

Improvements were made to the *Leiden jar* to increase its storage potential by lining the inner and outer surfaces with metal foil. The inside could be charged from a machine using a metal chain. The jar was a capacitor, which could condense weak static electric charges to give powerful shocks and could be used to carry charge from room to room.

Several jars could be linked in series to form *'batteries'* of increasing power. King Louis XV of France devised a shocking entertainment for his court; he made his guardsmen hold hands and then put a charge through them. This gave them a *'sudden spring'*. The entertainment was repeated with several hundred Carthusian monks. Electric shocks became the rage and therapeutic benefits were soon claimed.

One of the most important influences in the history of physiology is that of Count Albrecht von Haller (1708 – 77) the Swiss physician who studied under Boerhaave at Leiden. It is very likely that he would have been aware of the electrical experiments conducted at his old university. Haller became professor of anatomy, surgery and botany at the newly founded University of Göttingen and established the medical school there along the same principles as Leiden. He wrote an eight-volume work on physiology *Elementa Physiologiae* between 1757 and 1766 in which all contemporary physiology was discussed and all statements of a vague or mystical character were discarded. Haller supported well-conducted animal experiments to resolve doubts and based his physiological views on the principle that **irritability** is the property of muscles that **sensibility** is the property of nerves and that ultimately there would be a scientific explanation for nerve and muscle function.

FRANKLIN

If one man can be said to be responsible for America's love affair with inventions and gadgets, it must be Benjamin Franklin, publisher, inventor, scientist and statesman, who was born in Boston, Massachusetts in 1706, the fifteenth child of an English candle maker. As well as the *lightning rod* he invented *the Franklin stove, the medical urinary catheter, swim fins, the glass armonica and bifocal spectacles*. At the age of ten Franklin left school to help his father and was then a printer's apprentice. After spending two years in London where he worked as a compositor for a printer near St. Paul's cathedral, he returned to Philadelphia in 1730 and published the *Pennsylvania Gazette*. After some years he became a wealthy merchant.

In 1743 he helped found the American Philosophical Society for scientists to demonstrate their experiments and discoveries. Franklin learned of Gray's discoveries and embarked on his own research.

He proposed that *'vitreous electricity'*, produced by rubbing glass with silk like Gray and *'resinous electricity'* produced by rubbing amber with a duster like Gilbert, were not different types of *'electric fluid'* but the same fluid under different pressures. This was the *'one-fluid'* theory. Franklin succeeded in making Leiden jars, which he called Musschenbroek's *'wonderful bottles'*, store either positive or negative charges and noticed that pith balls with like charges repelled one another whereas opposite charges were attracted. Franklin also observed that if the wire and water inside a bottle are *'electrised positively* or *plus'* then the outer coating is simultaneously *' electrised negatively or minus in exact proportion.'* He was *"astonished at the wonderful way in which these two states of Electricity, the plus and the minus are combined and balanced in this miraculous bottle."*

Franklin realized that charge was stored on the surface of the glass of the Leiden jar, which was an insulator and designed a simple condenser or capacitor made of a flat glass plate separating two metal plates. Franklin noticed that the amount of charge stored was proportional to the area of the glass plate and made a *'battery'* of eleven alternating metal and glass plates connected in series to store considerable amounts of charge.

Of course unknown to Franklin, rubbing glass with silk removes electrons, to give *'vitreous electricity'* which he labelled *'positive'* and rubbing resins with a duster deposits electrons, to give *'resinous electricity'* which he called *'negative'*. Before the phenomenon of electric current was fully understood, the direction of electric flow was thought to be from positive to negative and known as *conventional current,* that is in the opposite direction to the flow of electrons, which when they were discovered, had to be assigned negative charge.Franklin published his book *Experiments and Observations on Electricity, Made at Philadelphia in America* in 1751, which became a standard work. He was awarded the Copley medal of the Royal Society and elected as a Fellow.

Franklin tried to understand how clouds charged themselves with lightning and suggested that the *'fire'* was generated by oceanic friction then held on the surface of clouds and discharged as a shock when a cloud met a tree, steeple or mast. In 1750 he performed experiments to prove that lightning is electric by flying kites into clouds in thundery weather and extracted some charge but these experiments were not written up until Priestley wrote his *History of Electricity* in 1767. Franklin's experiments led to the invention of the lightning conductor – *"upright Rods of Iron, made sharp as a Needle and gilt to prevent Rusting, and from the Foot of those Rods a Wire down the outside of the Building to the Ground;...Would not these pointed Rods draw the Electric Fire silently out of a Cloud before it became nigh enough to strike and secure us from that most sudden and terrible Mischief."* The lightning rod has prevented many fires and saved many lives since its invention. Like other electricians of his day, Franklin gave electric shocks in the treatment of paralysis but concluded that *"I never knew any advantage from electricity in palsies that was permanent."*

Although Franklin is remembered by most people as a statesman and for the part he played in the American Revolution and the foundation of the United States of America, his international reputation was acquired initially as a scientist. He helped to promote higher education and established colleges that became the University of Pennsylvania. He freed his own slaves and campaigned energetically for the abolition of slavery. As a diplomat and politician he was the American agent in London before the French Revolution and Minister to France during it. He alone of the Founding Fathers signed all three of the major documents, which he helped to write: *The Declaration of Independence, The Treaty of Paris* and the *Constitution.*

Franklin was probably responsible for writing the rights of authors, publishers and inventors into the Federal Constitution, which directed that the only private law topics to which the power of Congress extended were patents, copyright, bankruptcy, the value of money and commerce between the states. All other private law matters were the prerogative of individual states. Franklin was a huge man physically as well as intellectually. When he died aged 84, he weighed 300 pounds. His influence survives today and the principles of Human Rights and Intellectual Property Rights are still the concern of global politics and trade. He is also responsible for making electrons flow backwards.

In 1785, the French physicist and army engineer, Charles Coulomb (1737 – 98) devised a simple and efficient balance to study the laws of force between electrostatic charges applied to pith balls. He discovered that like charges repel and unlike charges attract with a force which is mathematically related to the distance between them. The unit of electric charge is named after Coulomb [1].

GALVANI AND VOLTA

Alessandro Giuseppe Volta (1745 – 1827) was a member of a religious family from Como. His uncles were all churchmen and his father was a Jesuit for eleven years before leaving the order to maintain the family line; however of his seven children all but Alessandro and one sister entered the Church. Volta became interested in electricity whilst at school, which was the first building in Como to have a Franklin rod installed.

Volta became a teacher at the Como Gymnasium where he experimented with Leiden jars and improved on them by using a flat-bottomed dish separated from an adjustable metal disc by a *'cake'* of wax and resin. These are stronger insulators than glass, so Volta's condenser retained more charge than a Leiden jar of the same dimensions; he called the capacitor an *electrophore*.

Volta observed that he could alter the electric capacity of his *electrophore* by altering the distance between the metal disc and the metal dish. When the two were close the capacity increased, when they were moved apart the capacity reduced but he observed with his electroscope that the tension of the charge increased correspondingly.Volta's electric tension is equivalent to electrical *potential difference* or *voltage*. The ***volt*** is the unit of electromotive force or potential difference.

In 1778 Volta was appointed professor of physics at Pavia University. In 1791 the professor of anatomy at the University of Bologna, Luigi Galvani (1737 – 98), published his study of the electrical excitation of disembodied frogs' legs. He explained the jerking of a leg upon completing a circuit through the nerve and leg muscle as a direct result of the discharge of a *'nerveo-electrical fluid'* previously accumulated in the muscle, which he supposed to act like a Leiden jar. Galvani thought the *'fluid'* was similar to frictional electricity.

When Volta learned of Galvani's experiments he dismissed them as *"unbelievable"*. In any case he was sceptical about physicians who he found to be *"ignorant of the known laws of electricity"* and he recognized *'animal electricity'* only in electric fish such as the sting ray. Possibly, the profusion of medical quacks peddling electric shocks to their patients coloured his view of physicians; another might have been his resentment that clinical professors were paid significantly more than physics professors. Volta was reluctant to investigate Galvani's claims, but was urged on by his medical colleagues at the university and so he planned and executed experiments that step by step brought him to the invention of the *electric pile or battery.*

Galvani's electric circuit was composed of two different metals; a brass skewer which attached the frog nerve to an iron frame. Volta was loath to dissect frogs so applied shocks from an *electrophore* and then from a bimetallic strip to a whole living frog and produced a similar trembling response. He then placed a piece of tin and a silver spoon on his tongue and experienced a most unpleasant taste. After many animal experiments Volta concluded that galvanic electricity arose from a difference between metals, that **electricity excited the nerve** and the **nerve then excited the muscle.**

Using the principle that combinations of different metals will produce electricity Volta designed a pile of alternating silver and zinc discs separated by damp cardboard, which provided continuous electric current when the circuit was completed between the top and bottom of the pile.

In 1800 Volta presented his discovery in a letter to the Royal Society as an *'artificial organ'. "Any one who touched both its extremities would enjoy the same sensation as grasping an electric fish."*

In the same year France retook the Milanese, which included Pavia, from Austria, Volta was invited to demonstrate his new electricity to the Paris Academy. Napoleon attended three sessions and awarded Volta a gold medal, a pension and made him a count and senator of the kingdom of Italy of which Napoleon was then king. The Austrians, who returned in 1814, imprisoned Volta for five years after which he retired to Como. The ***volt,*** the SI unit of *electromotive force* [***emf***]; or *potential difference,* was named after him [2]. The invention of Volta's pile meant that for the first time a continuous flow of electric current was available, in contrast to static electricity, which could only produce shocks of short duration. Experiments with flowing current quickly showed that some materials such as metals were good conductors of electricity and had low resistance to flow, that insulators had high resistance and many materials had intermediate electrical resistance.

In essence Volta's pile is equivalent to a modern battery with plates of different metals immersed in an electrolyte solution. The Voltaic pile was later improved using copper and zinc plates in a copper sulphate solution[3]. When such an electric battery is giving current, electrons pass in the circuit from the negative zinc terminal to the positive copper terminal; however the convention is that current flows in the opposite direction, from the positive terminal to the negative terminal (*conventional current*). So electrons flow backwards, in the opposite direction to conventional current.

ELECTROCHEMISTRY

When Volta's paper was presented to the Royal Society the possibility of using electric current from a battery to decompose water was discussed. Humphry Davy (1771 – 1829) who was still superintendent to Dr Thomas Beddoe's Pneumatic Institute in Clifton learned of this and he persuaded the Institute to provide him with a large voltaic battery made of over a hundred double plates for his experiments. Initially he failed to decompose pure water in his electrolytic cell; however he had success when acid was added. He was able to study the effects of electric current on different chemical solutions and seems to have been the first to realize that electricity was not generated by the mere contact of two different metals but depended on chemical reactions taking place.

In electrolytic cells, Davy proved that the applied current acted to separate chemical compounds into their components, rather than to synthesize new substances. In a cell there are two electrodes, the *anode* (positive electrode) and the *cathode* (negative electrode). Chemical ions that are attracted to the *anode* are known as *anions* and have negative charge, whereas ions attracted to the *cathode* are known as *cations* and have positive charge like hydrogen and metals.

In 1801 Davy was appointed lecturer, then professor of chemistry to the Royal Institution in central London and given a large amount of resources and a special laboratory for his electrical and chemical experiments. He also gave public demonstration-lectures, which attracted large audiences; similar public lectures have continued at the Institution up to the present day. Davy was elected Fellow of the Royal Society in 1803.

Davy finally succeeded in decomposing pure water into oxygen and hydrogen [4], and then in 1807 he discovered potassium by the electrolysis of potash solution in a cell. When the battery was switched on, globules of silvery matter collected at the cathode (negative pole) of the cell and oxygen was released at the anode. At first he called the matter *'potagen'* but realized that despite its extreme lightness it had the attributes of a metal and renamed it *'potasium'*. Davy repeated the experiment with soda as electrolyte and discovered sodium.

Using electrolytic methods he went on to isolate chlorine, calcium, magnesium and strontium, but from a biological view the discoveries of sodium, potassium, calcium and chloride ions are the most significant because of the crucial role these ions play in cell dynamics and particularly in the conduction of nerve impulses and muscle contraction.

His lasting fame was assured by the invention of the *Davy Safety Lamp* and the electric carbon arc, which was the first mode of electric lighting. The carbon arc is still used in arc welding. In 1812 Davy became the first scientist since Newton to be knighted for his many achievements in chemistry and electricity; his other great achievement was to appoint Michael Faraday (1791 – 1867) as his laboratory assistant.

NERVES

Just a few streets away from Davy's laboratory in the Royal Institution, was the Great Windmill School of Anatomy established in 1770 by the famous anatomist, gynaecologist and obstetrician, William Hunter (1718 – 83), in London's Soho district. Here Charles Bell (1774 – 1842) a Scottish surgeon and anatomist with an Edinburgh medical degree ran a medical school. He was also a military surgeon and in 1809 had treated many of the returned wounded from the Battle of Corunna. In 1815 when he was a surgeon to the Middlesex Hospital he left for Brussels immediately he heard of the Battle of Waterloo and reached the city ten days later. Bell operated for twelve hours each day until "*my clothes were stiff with blood and my arms powerless with the exertion of using my knife.*" None of Davy's nitrous oxide anaesthetic was available to relieve the soldiers' pain. Bell recorded many of the soldiers' wounds with vivid illustrations.

Charles Bell, like Thomas Willis more than a century before, performed animal experiments and made detailed descriptions and illustrations of his dissections of the human brain, spinal cord and peripheral nerves which have hardly been bettered and published *An Idea of a New Anatomy of the Brain* in 1809 in which he claimed that a nerve is either motor or sensory but not both, and also that stimulation of the anterior root of a mixed nerve arising from the spinal cord causes convulsions of muscles but stimulation of the posterior root did not. Bell also asserted that nerve impulses travelled only in one direction.

Bell's description of the cranial nerves supplying the head and neck divided them into nerves of special sense; the olfactory (smell), optic (vision), auditory (hearing) and lingual (taste); nerves of common sensation covering the forehead, face, and the throat; and nerves of motion supplying the muscles of the eye, face, tongue and pharynx. His experience as a war surgeon must have given him an understanding of the effects of nerve, spinal cord and brain injuries and their impact on sensory and motor function.

A controversy arose in 1822 when Francois Magendie (1783 – 1855) a French physician, claimed primacy in demonstrating that the anterior spinal root contained only motor nerve fibres and the posterior root only sensory fibres. Compromise was achieved by calling this the **Bell-Magendie Law.** After this there was an increasing realisation that sensory nerves transmitted electrical impulses of some kind but that the type of sensation depended on the sense receptor and not on the nerve. Newton had shown that stimulating the retina with either light or pressure from a bodkin gave a visual sensation of light; likewise Volta with an electric spoon and the sense of taste.

Like Thomas Willis, Bell combined the scientific study of neuro-anatomy with his clinical practice. He described the course of the facial nerve and the one-sided facial paralysis resulting from its disease or injury, which is still known as *Bell's Palsy.* He presented a paper to the Royal Society entitled *On the Nerves: Giving an Account of some Experiments on Their Structure and Functions* and was elected a Fellow in 1826 and was knighted in1831.

FARADAY

At the same time that Bell was investigating the structure and function of the human and animal nervous systems, Davy and Michael Faraday were continuing their chemical and electrical researches a few streets away at the Royal Institution. Faraday's father had been a blacksmith who became ill and unable to provide more than bare necessities for his family. Michael Faraday, who was born in south London, had little formal education and as a boy delivered papers for a bookseller to add to the family earnings. The bookseller then took him on as an apprentice bookbinder. Faraday enhanced his education by reading books as well as binding them, and came across the article on *'Electricity'* in *Encyclopaedia Britannica,* which questioned why conductors conduct electricity and insulators do not.

Faraday decided to study science and joined the City Philosophical Society. One of the bookseller's customers gave him tickets to attend Davy's lecture demonstrations in 1812 where he made meticulous notes. An explosion in his laboratory temporarily blinded Davy and Faraday was recommended as an amanuensis. On his recovery, Davy received the carefully bound notes and was so impressed that he appointed Faraday as a laboratory assistant in 1813. Faraday devoted himself to chemistry and established himself as an expert analytical chemist.

In 1820 the Danish scientist Hans Christian Oersted's discovery of a circular magnetic *'force'* around a wire carrying an electric current, prompted Faraday to experiment with electromagnetism.

Faraday found that an electric carbon arc deviated when a strong magnet was brought close and then Davy showed that iron filings formed up in circles on a bench with a current passing perpendicular to it, which indicated lines of magnetic force. Faraday established that a direct current induces a magnetic field in a plane perpendicular to the flow of current. Faraday then carried out a series of careful electromagnetic experiments using instruments of his own design, which gradually elucidated the relationships between electricity and magnetic forces, which were to lead to the invention of the electromagnet, the transformer, and ultimately to the dynamo and electric motor.

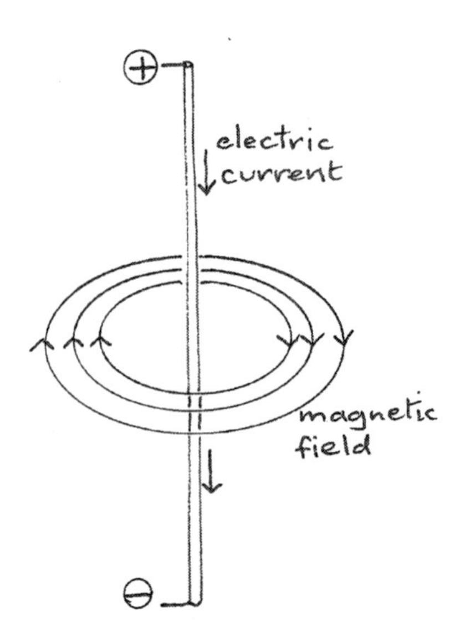

7. Faraday showed that a direct current induced a **magnetic field** in a **clockwise** direction, in a plane perpendicular to the current.

Faraday was not alone in researching electromagnetism and many groups in Germany, Denmark and especially Ampere in France were performing experiments. Despite Franklin's declaration of the *'one-fluid'* theory, most electricians in Europe still believed in *'two fluids'*, as an explanation for the properties of the electric battery. Faraday came to realize that Franklin had been correct and that the *'one-fluid'* concept was sufficient to account for electrical events.

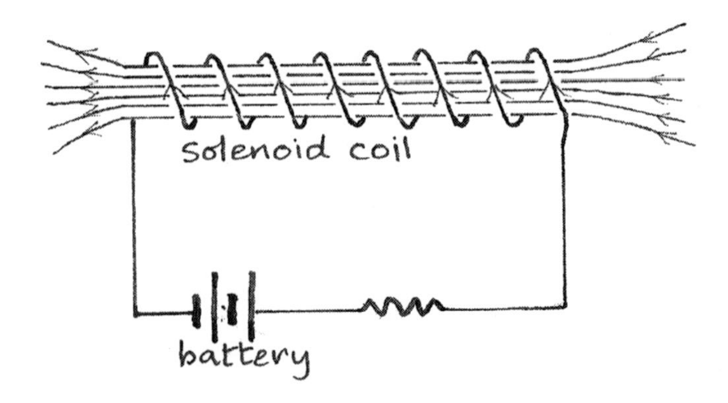

8. Faraday's solenoid electromagnet.

The ***electromagnet*** was invented by Faraday when he found that a helical wire coil or solenoid around an iron bar magnetized the bar when a current passed through the coil. Even when the iron bar was removed the solenoid still produced a magnetic field as current flowed through the circuit. He then devised an iron ring with a primary solenoid circuit on one side, then a secondary solenoid circuit around the opposite side and found that a current through the primary circuit induced a current in the second circuit. If the number of turns of the wire in both solenoids was the same, the voltage in the two circuits was the same but if the number of turns in the secondary solenoid was doubled the voltage in this circuit doubled. He had discovered the principle of the ***voltage transformer*** which could 'step-up' the voltage of the primary circuit.

These experiments showed how electric current could induce magnetization in ferromagnetic materials. Faraday then discovered that moving a magnet in relation to an electric circuit induced an electromotive force (*emf*), which produced a current in the circuit and conversely that moving a wire in a magnetic field also induced a current. This discovery subsequently led to the invention of the ***electric dynamo.***

ELECTRIFICATION

Faraday himself did not produce a serviceable magneto-generator although he did publicly demonstrate the induction of a current through a circuit with a copper disc rotated between the poles of the giant horseshoe magnet belonging to the Royal Society. His collaborator Charles Wheatstone (1802 – 75) and then Werner Siemens (1816 – 92) went on to devise useable electricity generators or dynamos.

Enthusiastic inventors caught the spirit of the age by transforming the newly acquired knowledge of physics into useful technologies. In 1877 Thomas Edison (1847 – 1931), an American telegraph operator invented a more effective transmitter for long distance electric telegraphy; then Werner von Siemens and his younger brother Sir William Siemens designed and laid some of the first overland and submarine telegraphic cables in Europe using their cable ship *Faraday*.

William Thomson, later Lord Kelvin (1824 – 1907), who was elected to the chair of natural philosophy in Glasgow at the age of only 22, clarified many theoretical problems in thermodynamics and electricity.

Kelvin also designed sensitive receivers to detect faint electric signals in long distance telegraph cables, which ensured their success.

In 1800 sending messages a long distance was limited by the speed of the courier or by the chain of signal towers; however by mid-century there were telegraphic links between many parts of Europe. The first transatlantic cable was laid in 1866 and by the end of the century there was a network of long distance cables providing telegraphic communication around the world.

At the beginning of the 19th century all lighting depended on oil lamps and candles. The introduction of town gas improved domestic and public illumination but still used bare flames. By mid-century Davy's electric arc lamp was used in large theatres and show-grounds but was inconvenient for domestic use.

In 1880 Edison in the United States and Swan in Britain invented the incandescent electric lamp using a carbon filament in an evacuated glass bulb. The light bulb became practical for domestic and public use when more durable filaments of osmium, tantalum or tungsten were used instead of carbon. The major problem was the supply of electricity. Small electric light companies used steam driven dynamos to supply electric current on a local basis, to perhaps a thousand or more lamps in nearby hotels and public buildings.

Direct current (DC) where the direction of current remains the same was easier to manage than alternating current (AC), where current changes direction many times a second; however with DC there is progressive loss of power with distance because of electrical resistance. DC power stations only have a range of a few kilometres.

One of the most important electrical engineers of his time, Nikola Tesla (1856 – 1943) was the younger son of a Serbian Orthodox priest in Croatia. As a child, Nikola's family blamed him for the accidental death of his clever elder brother. He suffered from chronic colitis and developed a severe obsessive/compulsive disorder, which persisted throughout his life. At school he became obsessed with electricity and attended Graz Polytechnic Institute where he studied electricity intensely. Tesla came to realize that AC power supply might have some potential advantages over DC.

Although it was more dangerous and difficult to use, AC could easily have its voltage stepped up or down by transformers and high voltage current might be transmitted for long distances with only slight loss of power. However no one had been able to make a practical AC system work because the constant change in direction caused violent vibrations in AC motors. Tesla realized that two coils, placed at right angles and supplied with alternating currents 90° out of phase with each other would create a rotating magnetic field, which could drive a motor. The vibrations caused by each of the two phases acting in opposition would cancel out so that AC motors could run smoothly.

Tesla got a job with the Continental Edison Company in France to fix the lighting and power for Strasbourg railway station and it was during this time he built his first AC motor and dynamo. In 1884 he used all his savings to sail to New York where he began working for Edison who had no faith in AC but was soon impressed by Tesla's dedication and ability. Edison promised Tesla a bonus if he could improve the efficiency of his DC motors, which Tesla did. Edison refused to hand over the money. Tesla protested unsuccessfully and quit. Tesla then made a series of business errors and was reduced to digging ditches for a dollar a day until George Westinghouse heard about Tesla's AC motors and dynamos. Westinghouse Electric Company employed him to build a system to generate electricity at Niagara Falls and to supply AC power to the city of Buffalo, 22 miles away; this was completed in 1895. Tesla's AC system was chosen to light the Chicago Exposition of 1893 with hundreds of thousands of light bulbs.

Tesla's AC electric power systems were widely adopted in Europe and North America because they could transmit large amounts of electric power from remote hydroelectric and coal fired generators to the hearts of large cities. Electricity could be used to power electric trams and trains, factories and hospitals as well as public and domestic lighting. Lines of electricity pylons striding to the horizon became symbols of modern civilization and development. So much so that in 1920 Lenin promised the backward Soviet republics electrification for the proletariat and by 1927 the Central Electricity Board in Britain started to build a national electricity grid to link all supplies and provide distribution and service to customers throughout the United Kingdom.

Tesla had powered the twentieth century. He went on to design a high frequency tuning coil, which Marconi used to make his first radio transmissions. Yet this tormented genius was almost completely forgotten, his memory overshadowed by Edison and Marconi who both took advantage of him. Tesla died alone and impoverished in a New York hotel room in 1943; he had probably been dead for two days before his body was found. Nine months later the US Supreme Court found that there was nothing in Marconi's radio invention that had not already been patented by Tesla.

At last this genius with an intuitive grasp of electromagnetic induction was recognized when the SI unit of magnetic flux density (proportional to magnetic field strength) was named after him [5].

ELECTRONS

When a separate metal plate was put into a light bulb, Edison showed that a current passed from the filament, through vacuum to the plate, but only when it was connected to the positive terminal of his lamp and not when it was connected to the negative terminal. With characteristic modesty he called this phenomenon, the *Edison Effect*. This discovery led to the invention of the cathode ray tube and also the thermionic electric valve, the fore-runner of the diode and the transistor.

A cathode ray tube is a vacuum tube with a heated filament cathode at one end and an anode at the other. When a strong *emf* is applied, greenish rays spread out from the cathode towards the anode, which were initially thought to be rays of light.

In 1897 Professor Joseph John 'JJ' Thomson (1856 – 1940) at the Cavendish Laboratory in Cambridge showed that the *'rays'* could be deflected by magnetic fields and were not rays but particles with mass and electric charge. Their mass was extremely small; each particle had about one thousandth the mass of an atom. He called the particles *electrons*. It became evident that good electrical conductors, such as metals, had mobile *electrons* in their atoms, which provided electric current when *emf* was applied and that electric current is a flow of *electrons*.

NEUROMUSCULAR ELECTRICITY

The two types of electric current have different biological effects, which were soon discovered in the mid 19th century when physicians used Galvanic current (direct) from batteries or Faradic current (alternating) from magnetos, rather than Franklinic (static) electricity, for treating their patients.

Writing in 1885, Vivian Poore, a physician at University College Hospital, London emphasized the dangers of constant direct, or galvanic, current which produced chemical effects and heating at the negative applicator (cathode) that could lead to inflammation or even sloughing of the skin. He wrote-

"Since the human body consists of a mass of cells which contain and are bathed in saline fluids, many of the phenomena observed on passing galvanic currents through the body are probably due to electrolytic action." Of alternating or Faradic current he wrote *"Its direction is constantly changing so that in using it, it is less necessary to distinguish between the poles."*

Poore noted that galvanic current only produced muscle contractions when it was switched on or off, whereas alternating current gave continuous muscle contractions.

Poore also recorded that galvanic current, unlike alternating current stimulated the nerves of special sense.

"If applied near the eyes flashes of light are seen, and blindness has resulted from the incautious application of strong currents to the face."

Electrical stimulation was used for neurological diagnosis to distinguish different types of paralysis but like Franklin, Poore found electrical treatment of palsies gave generally poor results. However he found it useful for the relief of neuralgias, headaches and sciatica. Writer's cramp and wryneck responded quite well too.

It was by this time clear that nerve conduction and muscle contraction were bioelectrical phenomena but how nerves transmitted electrical signals was still mysterious. Analogies with the electric telegraph were inadequate because nerve tissue was no better an electric conductor than most other body tissues.

For two hundred years after the invention of the compound microscope in 1610 no real improvement in its optical system had been achieved because of chromatic and spherical aberration. However in 1830 arrangements of achromatic lenses made of special glass led to significant improvements in microscopy. New tissue staining techniques revealed the **microanatomy** of nerve, muscle and also of the sense receptors in the skin, tongue, nose, inner ear and retina in greater detail than ever before.

Jan Evangelista Purkinje (1787 – 1869) grew up in Bohemia and entered a monastery to study for the priesthood. Just before his ordination he decided instead to study philosophy, including physics at Prague University. He transferred to medicine and graduated in 1819.

Purkinje became professor of pathology and physiology at Breslau University in Prussia from 1823 until 1850 when he returned to Prague as professor of physiology. He was a friend of Goethe and was the first to recognize fingerprints as a means of individual identification. He was most interested in the physiology of the heart and the brain and performed detailed microscopic studies in order to learn how they functioned. He was the first to use a microtome and tissue fixation to make very fine slices of muscle and brain tissue that had been stained with special dyes. The thin slice technique allowed the cellular structure of tissues to be observed with transmitted rather than reflected light and was more informative.

Muscles are of three kinds, skeletal, visceral and cardiac. **Skeletal muscle**, responsible for locomotion and potentially under voluntary control, accounts for 40 percent of human body weight. Under the microscope it is seen to be composed of fibres measuring 10 – 100 microns in width (a micron is one millionth of a metre) and 30 – 100 mm. in length. Microscopically, bundles of skeletal muscle fibres show transverse striations so that skeletal or voluntary muscle can also be called striated muscle. Each muscle fibre receives a nerve fibre, which terminates at a neuromuscular endplate where a nerve impulse stimulates muscle excitation and contraction by releasing the chemical neurotransmitter, acetyl-choline. Skeletal fibres each contract separately; however as more fibres within a muscle unit are stimulated, contraction becomes stronger.

Claude Bernard studied **neuromuscular transmission** using electrical stimulation. A friend provided him with some curare tipped arrows that he had been given by South American natives and which paralysed a rabbit in 6 minutes. Bernard found that electrical stimulation of a motor nerve in a curarized frog had no effect, whereas its muscles contracted when directly stimulated. This convinced him that curare blocks the nerve terminals (endplates) in muscles. It does so by blocking the action of acetyl-choline.

One of the consequences of the electrical revolution was the development of increasingly sensitive galvanometers to measure current. The French-Swiss professor of physiology in the University of Berlin, Emil du Bois Raymond (1818 – 1896) found, using a galvanometer, that currents detected during muscle contraction in the limbs of a frog could be found in the limb of any live animal and that the current of the whole limb is the sum of the currents of the individual muscles.

Although the action potentials of muscle and nerve fibres are similar, the large bulk of skeletal muscle generates much larger currents that are more easily detected and measured than nerve currents. German universities took the lead in electrophysiological research. Gradually the assumption that the phenomena of living animals depended on special biological laws and vital forces, was rejected in favour of explanations of physiological phenomena based on the very same physical and chemical laws that operate for inorganic matter.

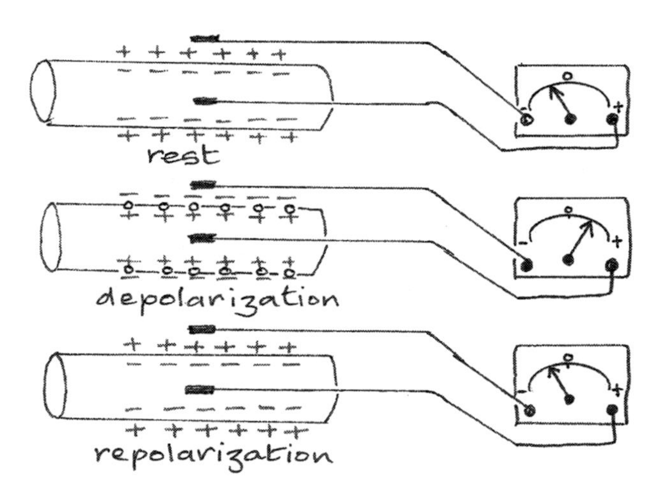

9. Electrical changes during muscle contraction. At rest the external surface of a muscle fibre is **positively charged** or **polarized** compared to the inside cytoplasm which has a **negative resting potential.** When stimulated, the fibre is **depolarized,** the external positive charge becomes negative and the **action potential inside the fibre becomes positive.** During **repolarization** the **negative resting potential** is restored.

The mechanism by which muscle current is generated is based on **depolarisation** of muscle fibres. The external surface of a **resting muscle fibre is positively charged** or **polarized** relative to its cytoplasm. When stimulated at the endplate there is loss of polarization, equivalent to an electrical short circuit across the cell membrane, and a **negatively charged wave of depolarisation** extends along the fibre stimulating contraction of the muscle proteins as it passes; **polarization** is restored after an interval of several microseconds.

The speed of conduction of the muscle wave is much slower than nerve conduction and measures only 3 to 5 metres per second. Electromyography is a method of measuring voltage changes during muscle contractions by inserting needle electrodes at separate points in a muscle; voltages generated are about 50 micro-volts. After motor nerve injury, de-nervated muscles produce spontaneous irregular contractions, several times a second, known as either *fasciculation* or *fibrillation*. Electromyography can be used to measure these phenomena and monitor nerve recovery.

Smooth or involuntary muscle in the muscular coats of the gut produces peristaltic movements which propel bowel contents. Smooth muscle in the walls of blood vessels or bronchi controls their calibre and the flow rates of blood or air. Microscopically, smooth muscle fibres are interconnected and do not show striations. When fibres are stimulated excitation spreads from fibre to fibre until the whole tissue contracts. In smooth muscle, excitations may arise rhythmically within the muscle itself and this activity can be inhibited or enhanced by nerve impulses originating in the autonomic (involuntary) nervous system or by hormones.

THE HEART

Cardiac muscle shares some characteristics of both skeletal and smooth muscle; like smooth muscle it is involuntary, its cells are interconnected and it displays intrinsic contractility and rhythmicity that accounts for the heart beat. Like skeletal muscle it conducts its action potentials in a linear and coordinated manner and it is striated but much less so than skeletal muscle. Jan Purkinje was aware of the pattern of normal contraction of the muscle in the walls of the cardiac atria and ventricles and which comprise the normal cardiac cycle of systole and diastole.

Systole is when cardiac chambers contract to pump blood and **diastole** is when they relax and refill. Although the heart is supplied by autonomic nerves that can modify the heart rate, a de-nervated heart will continue beating regularly, as after cardiac transplantation. The cardiac muscle impulse starts in the right atrium, in a focus called the **sino-atrial node**, spreads diffusely through both atria, causing atrial systole, then to a node or focus called the **atrioventricular node (AV node)**.

In 1839 Purkinje described the bundles of special conducting fibres, which transmitted the cardiac impulse from the AV node then throughout the muscle of both ventricles. These Purkinje fibres are not nerves but modified muscle fibres, which transmit the impulse at an appropriate rate for orderly contraction of the ventricles.

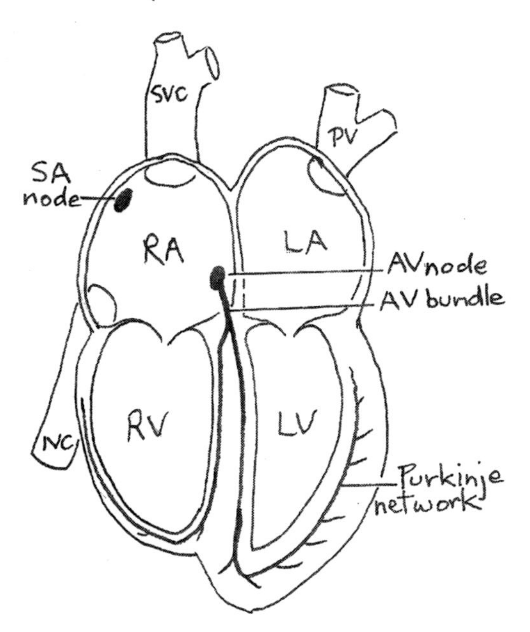

10. Electrical Conduction through the Heart. The cardiac impulse is wave of depolarization which starts at the **sino-atrial node (SA node)**, spreads to the **atrio-ventricular (AV) node AV bundle** and **Purkinje network** of the myocardium. It is followed by a recovery wave of repolarization. **SVC & IVC**, superior and inferior vena cava; **RA&LA**, right and left atria; **RV&LV**,right and left ventricles; **PV**, pulmonary veins.

Blood returning from the major veins into the atria mostly passes into the ventricles before atrial systole, which assists ventricular filling but muscle in the atrial walls is relatively thin. Atrial systole precedes the more powerful ventricular systole, which suddenly closes the atrio-ventricular valves and ejects blood at high systolic pressure through the opening pulmonary and aortic valves. The muscle wall of the left ventricle is much thicker than of the right ventricle because the systemic circulation is more extensive than the pulmonary circuit and aortic pressure much higher than in the pulmonary artery. When ventricular systole is complete, the ventricles are emptied and the pressures in them fall so that the aortic and pulmonary valves close abruptly.

When listening to the heart with a stethoscope, the '*lub-dup*' sound is caused by closure of the valves by sudden pressure changes; '*lub*' marks closure of the AV valves at the start of systole and '*dup*' the closure of the aortic and pulmonary valves at the end.

In 1896 electrophysiological techniques, pioneered in the Berlin Institute of Physiology to study skeletal muscle, were applied to heart muscle using electrodes placed on the surface of the chest; changes in current were detected by a galvanometer. In 1903 Eintoven described electrical tracings from the heart using a triangular array of electrodes and a sensitive galvanometer consisting of a silver-coated quartz fibre suspended between the poles of an electromagnet. The tracing was called an **electrocardiogram (ECG).** Modern ECG machines use amplifiers and oscilloscopes and are much more sensitive than Einthoven's galvanometer. They are available in medical clinics to assess the effects of coronary artery disease, heart valve disorders and hypertension on cardiac muscle (myocardium), as well as to assess abnormal heart rhythms or arrhythmias.

The ECG of a healthy person presents a complex pattern of waves; **P,Q,R,S** and **T**, which reflects transmission of the depolarizing cardiac impulse from the **sinoatrial node** to the **atrioventricular node**, then throughout both ventricles, followed by **repolarization** over each cardiac cycle.

The first small wave **P**, represents **atrial depolarisation;** the large triphasic complex, **QRS, ventricular depolarization**; and **T, ventricular repolarization** or **recovery.** The QRS complex is triphasic because current flows in different directions around the right and left ventricles in relation to the recording electrodes.

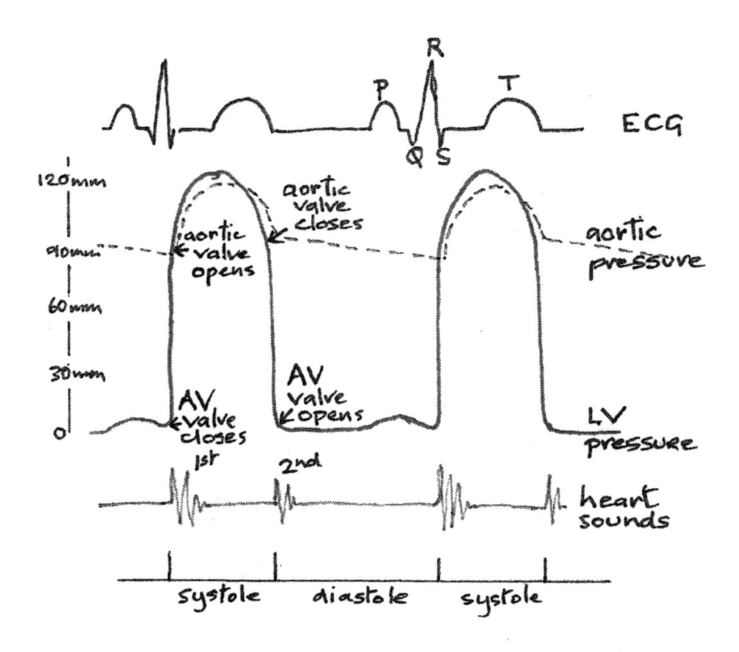

11. Electrocardiogram (ECG), aortic and left ventricular (LV) blood pressures during the **cardiac cycle.** Rising LV pressure at the onset of ventricular systole closes the AV valve and opens the aortic valve. At the end of systole the aortic valve closes and the AV valve opens.

A common arrhythmia, which may occur spontaneously, in people with overactive thyroid glands or known heart disease, is **atrial fibrillation (AF).** The **sinoatrial** node is normally the '*leader of the band*' and starts each cardiac cycle, however if abnormal foci start irregular excitations the atria do not contract effectively but fibrillate or quiver. The loss of atrial propulsive power in AF is not so very important; the main problem is the irregular rate at which cardiac impulses reach the **AV node,** which excite the Purkinje system unpredictably and which causes ventricular systole to occur before time and before the ventricles can fill properly with blood. No P waves are seen on the ECG and the QRS complex is fast and irregular; the patient experiences palpitations and may go into heart failure.

Atrial fibrillation can usually be managed effectively and is unlikely to be fatal, whereas **ventricular fibrillation** (VF) following sudden blockage of coronary arteries or electrocution is more often fatal, because the ventricular pump fails and no fresh blood reaches the heart muscle or the brain.

In cardiac arrest the patient is unconscious, pulse-less and not breathing. Irreversible brain damage occurs within three minutes if an adequate circulation is not re-established. Immediate **cardiopulmonary resuscitation (CPR),** involves rapid sternal compression of the ventricles, mouth-to-mouth ventilation and defibrillation as soon as possible.

An *alternating* current of 50 or 60 cycles per second will repeatedly depolarise the heart and induce ventricular fibrillation, which is usually the cause of death by electrocution. It was for this reason that Edison was wary of AC. Tesla taught his engineers always to keep one hand in a pocket when dealing with high AC voltages so that an unexpected shock would pass from one hand to the feet; not between the hands and through the heart. This advice undoubtedly saved many engineers' lives.

In the presence of **VF** the passage of an intense *direct* current through the heart of several thousand volts for a few microseconds with a **cardiac defibrillator** may depolarise all the cardiac muscle simultaneously. After a refractory period the **SA node** may once again take the lead and establish a normal cardiac rhythm.

However muscle damaged by myocardial infarction, which caused VF initially, is likely to fibrillate again. A cardiac defibrillator has a large capacitor that can discharge 200 to 360 kilojoules through two chest electrodes, one over the right atrium and the other near the apex of the left ventricle. Care must be taken to avoid giving rescuers an electric shock when resuscitating a patient with VF.

ACTIN AND MYOSIN

The complex electrochemical mechanism, which causes contraction of muscle fibres was not understood until after 1950 when the combined efforts of biochemists, electron microscopists and electro-physiologists elucidated the *'sliding filament hypothesis'* involving the muscle proteins **actin** and **myosin**. Much of the collaborative research was performed at the Woods Hole Research Institute, Massachusetts, which was directed by Albert Szent-Györgyi (1893 – 1986) the Hungarian biochemist, who had received the Nobel Prize for identifying vitamin C in 1937 and who wrote *The Chemistry of Muscle Contraction* in 1951. He later wrote that most of the preceding theories of muscle contraction were "*fatally hurt in the impact between physics, chemistry, physiology and electron microscopy, fitting only the science of their author but being incompatible with that of others.*"

Two researchers working independently and by strange coincidence sharing the same surname though not related, Andrew Huxley and Hugh Huxley, played major roles in proposing the elegant '*sliding filament model*' of muscle contraction.

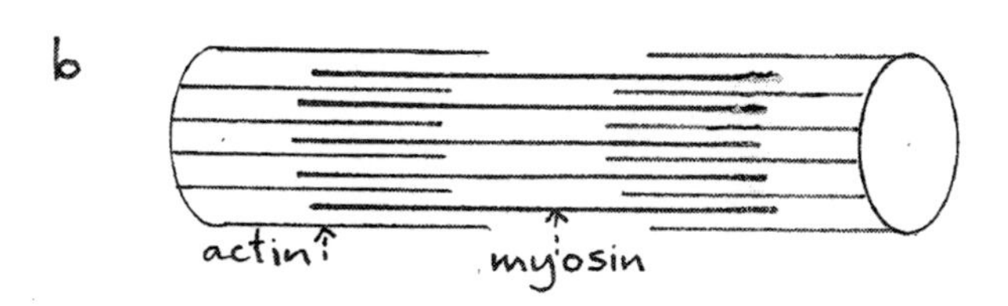

12. **Muscle fibres** are made up of bundles of **myofibrils** (**a**), which are in turn, made of **columns of sarcomeres** (**b**) composed of interdigitating protein filaments of **actin and myosin**.

Examination of the smallest elements of a muscle fibre; **myofibrils,** with electron microscopy and x-ray diffraction techniques, reveals that **myofibrils** are long columns of units called **sarcomeres** and that these are composed of alternating filaments of **actin** and **myosin** which are interleaved or interdigitated.

When a muscle fibre is stimulated it becomes **depolarised** and its membrane is short-circuited so that **calcium ions (Ca^{++}),** which at rest are kept outside the fibre, surge into the fibrils causing **actin** and **myosin filaments** to **interdigitate** shortening the sarcomeres and the myofibrils, so causing **contraction** of the muscle fibre. The contraction is powered by **ATP,** which is also used up ejecting calcium ions out of the fibre during **repolarization.**

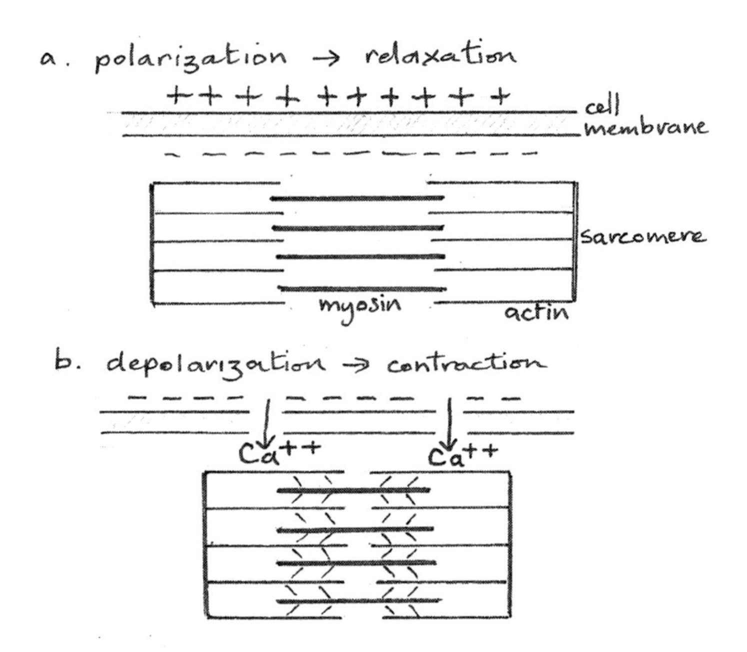

13. Sliding Filaments. a. At **rest** the outer membrane of the myofibril is **positively charged** relative to its cytoplasm; **actin** and **myosin** filaments are separated. **b.**An **excitation wave** causes **depolarization** of the membrane and **calcium ions** (**Ca**$^{++}$) flood into the cell attracting myosin to actin, shortening the myofibrils and causing **muscle contraction.**

NERVE IMPULSES

Von Helmholtz, whilst still a professor of physiology, devised a method to measure the time elapsed between electrical stimulation of a nerve and consequent muscle contraction, using a specially designed galvanometer. For the leg nerve of a large frog he calculated quite accurately the speed of transmission, at room temperature (20°C) to be 25 to 40 metres per second. For human myelinated nerve the speed can vary between 30 to100 metres per second.

By the end of the 19th century it was well established that a rapid wave, the nerve impulse, could be set up in nerves by electrical stimulation, however it was difficult at that time to measure the changes because they were so small and so rapid. Although the cathode ray oscilloscope was invented in the 1900s it was insufficiently sensitive to measure small voltage changes. However by 1925 oscilloscopes and amplifiers, designed for telegraphy, were able to detect small signals arising from nerves.

Edgar Douglas Adrian (1889 – 1977), (later Lord Adrian) qualified in medicine at Cambridge in 1915, where he became a clinical neurologist and then professor of physiology. He studied the action potentials of anaesthetized cat, rabbit and human nerves using a triode amplifier and oscilloscope, which magnified recorded responses by nearly 2000 times.

Adrian established the *'all or none'* principle of stimulated nerves and recognised that the passage of a nerve impulse leaves in its wake a refractory period during which the excitability of the nerve is depressed but then after a brief interval returns to normal.

However the mechanism for transmission of the nerve impulse was still unclear. Adrian wrote *The Mechanism of Nervous Action* in 1932 and in the same year was awarded the Nobel Prize for Physiology or Medicine jointly with Sir Charles Sherrington.

Between 1934 and 1954 two neuroscientists working in Adrian's department in Cambridge and at the Marine Biology Laboratory in Plymouth, Alan Hodgkin and Andrew Huxley, studied the giant nerve axon of the Atlantic squid, which enabled them to record ionic currents as they would have been unable to do with almost any other neurone, because most neurones were too small to be studied by the techniques available at that time. Their experiments in Plymouth were interrupted by World War II, when Hodgkin went to work on radar but they were able to resume their research after the war.

With great skill, Hodgkin and Huxley inserted microelectrodes inside the squid neurones and were able to measure the cell **membrane potential** between the interior and exterior of the nerve fibres. They discovered that at rest the exterior of the nerve fibre was positively charged compared to the interior and called this the **resting membrane potential.** When the fibre was stimulated, the passage of the nerve impulse was associated with a reversal of polarity, or **action potential** so that the exterior became negatively charged compared to the interior for a brief moment of **depolarization.** Passage of the **action potential** was quickly followed by **restoration of the resting membrane potential**, or **repolarization.** They also studied the movement of **sodium** (Na^+) and **potassium** (K^+) **ions** during nervous activity because they suspected that movement of these ions across the cell membrane would account for the electrical changes they were measuring[6].

An **action potential** can be elicited in a nerve fibre by electrical, chemical or mechanical stimulation, which disturbs the equilibrium and suddenly increases the permeability of the membrane to sodium ions, which rush inside to depolarise the membrane and propagate the action potential along the nerve fibre.

Depolarization represents an electrical short-circuit across the membrane and reverses the membrane potential. In other words the exterior of the nerve fibre becomes electronegative, equivalent to a surge of electrons out of the fibre to match the rush of sodium ions into it. Very rapidly the membrane closes to sodium ions and **repolarization** then restores the resting potential once again, after a brief refractory period of a few milliseconds. The energy for maintaining the resting potential is provided by **ATP,** which drives the **sodium/potassium 'pump'.**

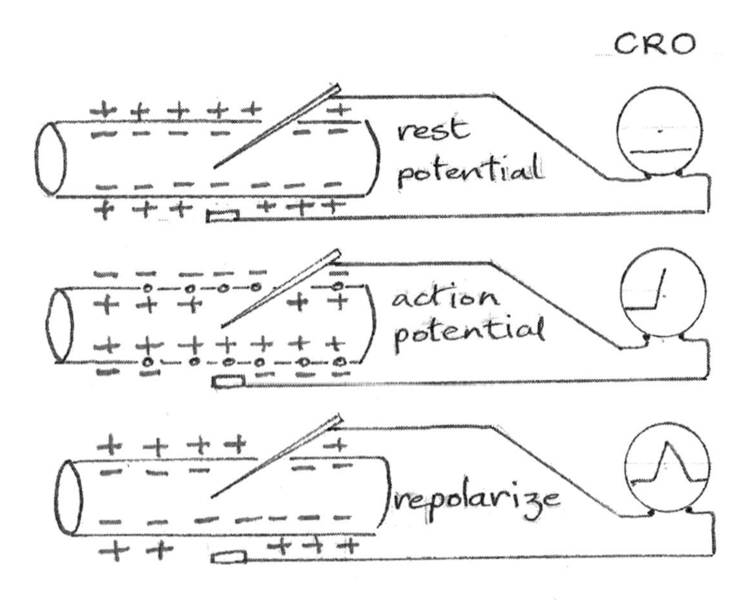

14. Microelectrodes inside and on the external surface of a **nerve axon** monitor millivolt changes during passage of the **action potential.** Signals from the nerve electrodes are amplified and displayed on a **cathode ray oscilloscope** (CRO).

The theory of nerve action potentials and the hypothesis of the existence of ion channels was published in 1952, but the ion transport hypothesis was not confirmed until later. Huxley the physiologist and Hodgkin the biophysicist shared the Nobel Prize for Physiology or Medicine in 1963.

As we have already seen, Andrew Huxley continued his research with muscle and found the importance of a similar mechanism for the contraction of muscle cell proteins.

THE BRAIN AND NERVOUS SYSTEM

In the early 19th century the anatomy of the brain, as revealed by dissection had been well documented. The first edition of *Henry Gray's Anatomy* appeared in 1858 and its description of the morphology of the human nervous system was comprehensive, however the way the brain performed its functions was poorly understood.Research then advanced on four major fronts: microscopic, clinical, electrical and pharmacological.

Microscopic studies using instruments with achromatic lenses made significant strides. Jan Purkinje gave an early description in 1837 of nerve cells and fibres in the brain; the cells in the cerebellar cortex, which he had studied, were named after him. Because of the limitations of his microscope, he thought that the nerve fibres were connected together in a continuous network.

Camillo Golgi (1852 – 1934) and Santiago Ramon y Cajal (1843 – 1926) met only once, when they were jointly awarded the Nobel Prize for Physiology or Medicine for their detailed morphological characterization of neurones and efforts to correlate structure with function. Like Purkinje they were restricted by the poor resolution of their light microscopes but Golgi then developed a silver stain, which allowed neurones to be seen more clearly and which Cajal refined for use in his own research.

Golgi and Cajal confirmed that the **nerve cell** or **neurone** is the basic structural unit of the nervous system and consists of a cell body with one or more projecting fibres, which vary widely in length and thickness. Multiple, branching, short fibres which transfer a nerve impulse toward the cell body are called **dendrites.** Single fibres that transmit an impulse away from the cell body called **axons** can be short and branching like dendrites or extremely long, measuring up to a metre or more. A motor nerve fibre can extend from its cell body in the cerebral cortex to the lower spinal cord, where its terminal fibres contact the dendrites or cell body of another neurone, whose axon then travels via the sciatic nerve to the forefoot. Just two nerve cells from top to toe! As Bell and Magendie had shown for nerve roots, individual nerve fibres can be either sensory or motor; and they can be either myelinated; meaning that the axon is surrounded by a myelin sheath, or unmyelinated. **Myelin** is a phospholipid substance which acts as an insulator and which is white in colour as opposed to the greyish colour of nerve cell bodies.

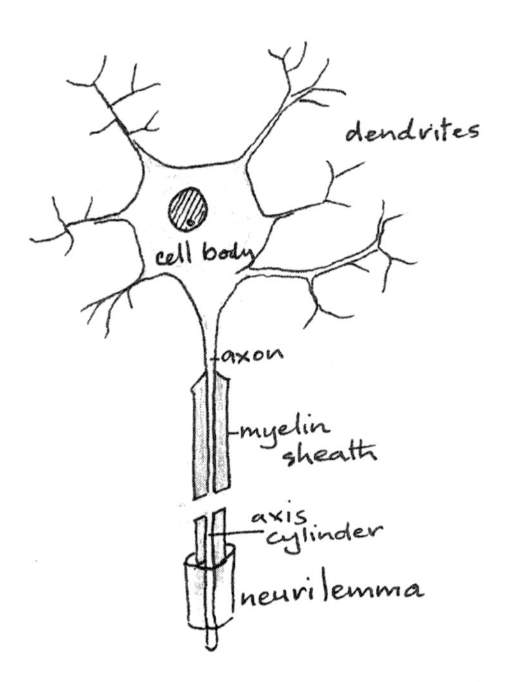

15. A Neurone with **dendrites** that conduct nerve impulses toward the cell body and a **myelinated axon** that conducts impulses away from the cell body. Myelinated axons in peripheral nerves are protected by a **neurilemmal sheath**. After Cajal.

The two Nobel laureates differed on one vital issue; Golgi saw no reason to dispute the long held view that the cells of the brain formed a continuous network whereas in the 1880s Cajal working in almost complete scientific isolation in Spain had realized that individual neurones were discrete entities.

Neurones are linked together in the nervous system, axon to dendrite or to nerve body, at points of contact which were later called **synapses** by Charles Sherrington in 1897.

At a **synapse** the nerve cells are separate but are so very close that a chemical **neurotransmitter** released by an axon, such as **acetyl-choline,** can stimulate adjacent cells. It was Cajal who postulated that the nerve impulses pass in dendrites towards the cell body and in the axon away from the cell body.

The anatomy of the human brain, spinal cord and peripheral nervous system is astoundingly complex. It has been estimated there are about 100 billion neurones in a healthy adult human brain, equivalent to the number of trees in the Amazon rain forest. However on dissection it shows regions mainly of two kinds comprised of either grey matter or white matter. The **grey matter** is composed of numerous nerve cell bodies, their interconnected dendrites, short axons, and supporting cells called **neuroglia.** It is rich in synapses and very active metabolically, requiring a constant supply of glucose and oxygen to function, so that it is highly vascular. **White matter** is composed predominantly of myelinated nerve axons, which connect the different regions of grey matter within the brain and spinal cord; it is less vascular than grey matter but also has supporting neuroglial cells.

The human nervous system has three functional levels –

higher brain or cortical level
lower brain
spinal cord and peripheral nerves

Cortex . The most striking grey matter region of the human brain is the folded outer covering of the cerebral hemispheres called the **cerebral cortex,** containing six layers of cells and measuring between 2mm. and 5mm. in thickness. The area of the cortex, if it were unfolded is about a quarter of a square metre; the bulges are called convolutions or gyri.

The two largest folds or fissures on each side are between the temporal and parietal lobes of the cerebrum - the Sylvian fissure, and between the frontal and parietal lobes - the fissure of Rolando, which is important because the **motor cortex** lies in front and the **sensory cortex** behind this fissure.

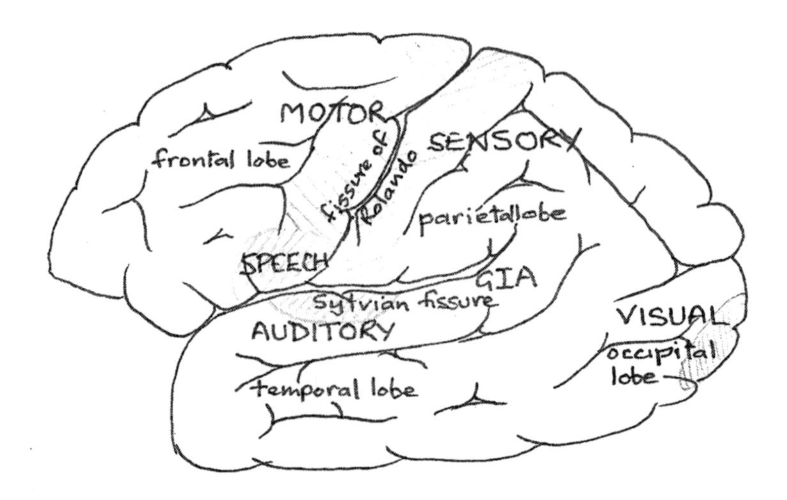

16. Side view of the left cerebral cortex showing areas associated with **vision**, bodily sensation (**sensory**), voluntary movement (**motor**) and **speech.** These **association areas** surround and feed into the **general interpretive area (GIA).**

Lower brain. Positioned centrally within the cerebrum and connected by white matter tracts to the cortex above, the brain stem, cerebellum and cord below, are the **thalamus** and **basal ganglia**. The **cerebellum** also has a cortex, similar to but less complex than the cerebral cortex; it is packed more neatly in parallel folds called folia; it is responsible for unconscious control of movement, balance and posture. Multiple centres of grey matter are present in the brain stem surrounded by white matter tracts and a central column of grey matter. These centres in the brain stem are responsible for automatic control of basic functions such as blood pressure and respiration.

Spinal cord. This is composed of *ascending and descending columns of white matter fibres* on the outer aspect of the cord, which connect centres in the brain stem, basal ganglia and cortex to neurones at different levels, and *grey matter* in the centre of the spinal cord. The *cord grey matter* has an 'H' shaped configuration in cross-section and is composed of cell bodies, dendrites and synapses of motor and sensory neurones whose nerve fibres extend either upwards in the cord or into the **peripheral nerve roots,** where they are bundled and protected by a sheath of connective tissue.

The neurologist John Hughlings Jackson (1835 – 1911) trained in medicine at York and then became physician to the London Hospital and the National Hospital for the Paralysed and Epileptic in Queen's Square. In 1863, he described the focal fits starting in the hand, then spreading to the arm and the rest of the body which are still known as '*Jacksonian Siezures*'. He associated certain areas of the cortex with limb movements and found that aphasia (inability to speak) was usually due to lesions on the left side.

Jackson proposed the principle of evolving or increasing complexity of the nervous system from automatic centres in the spinal cord and brain stem, through the more highly organized motor areas of the cortex to the most complex areas in the prefrontal cortex. At about the same time the Parisian neurologist and surgeon Pierre Broca (1824 – 1898) discovered that the power of speech was localized to a small area of the dominant cortex; the left side in right-handed individuals.

In the 1870s, as a consequence of the neurological discoveries of Hughlings Jackson in London, and of Charcot and Brown-Sequard in Paris, laboratory experiments on monkey brains were performed at Queen's Square, by the neurophysiologist and pioneering surgeon Victor Horsley (1857 – 1916).

Horsely studied medicine in London and neurophysiology in Berlin with du Bois Raymond. He returned to University College Hospital, London as a house surgeon in 1881 and was appointed surgeon to the National Hospital in Queen's Square in 1886 where he was a colleague of Hughlings Jackson and of Vivian Poore, the electric physician.

At Queen's Square, Horsley carried out studies of the functions of the brains of animals and humans. In the laboratory he studied the motor response to Faradic stimulation of the cerebral cortex, internal capsule and spinal cord. These studies were translated into his pioneering work on neurosurgical treatment for epilepsy and in 1884 he was the first surgeon to use intra-operative electrical stimulation of the human cortex for the localization of epileptic foci. Horsley mapped the important motor areas in the human brain and noted the similarity to those of monkeys. He was also the first surgeon to successfully remove a spinal cord tumour.

Horsley wrote *Functions of the Marginal Convolutions* in 1884 and *Functions of the Cerebral Cortex* in 1888. This remarkable man also pioneered carotid artery ligation for bleeding cerebral artery aneurysms and stereotactic surgery, for which a set of precise numerical coordinates were used to locate each brain structure in relation to the surface of the cranium. He also defined thyroid deficiency as the cause of myxoedema and treated it with monkey thyroid extracts. Horsley was a strong advocate of women's suffrage and of National Health Insurance to provide medical care for the needy. He was knighted in 1902. In 1916 Horsley died on active service as a British army surgeon in Iraq.

German neurophysiologists in the mid 19th century were much concerned with the integration by the brain of sensory input with motor action, particularly the perception of surrounding objects and their spatial relationships. The *nativistic* school thought that sensory-motor integration was *'hard-wired'* in the nervous system, whereas Helmholtz was an empiricist who believed that the brain constructed hypotheses about objects from vision, sound or touch, then acted by focusing the eyes, tuning the ears or adjusting finger movements, then reappraised sensory input in a continuous and integrated **'feed-back'** or **servo mechanism.**

In his 1855 Kant Lecture, Helmholtz asked *"How is it we escape from the world of the sensations of our nervous system into the world of real things?"*

He answered this philosophical question with scientific realism. We get outside the world of our retinas to the real space about us by practical action. Sensory *'feed-back'*, following eye or body movement builds up a mental construct of objects and movements in our environment.

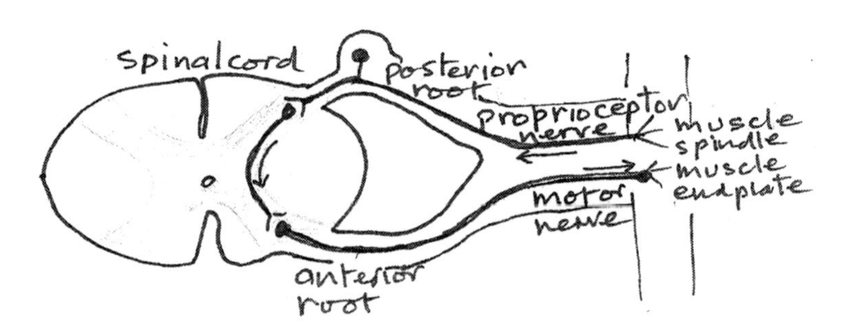

17. The Spinal Reflex Arc. The motor nerve with its cell body in the anterior horn of the spinal cord grey matter is the **final common pathway** for **motor impulses. Proprioceptive impulses** from pressure receptors in muscle and tendon in response to motor action, return to posterior horn, then through synapses with connector and motor neurones completing the **reflex arc** which acts as a **servo-mechanism.**

A **nervous reflex** is an involuntary act initiated by a sensory stimulus, which provokes a muscular response; for example rapid closure of the eyelid when an object touches the eyelashes; withdrawal of the hand from a hot object; recovery of balance after a slip or the knee-jerk response. We are conscious of such reflex actions only after they have occurred because they are mediated by reflex arcs, which at their simplest consist of two neurones linked together in the nervous system, axon to dendrite in the spinal cord.

A receptor neurone transmits a sensory impulse from the skin or pressure sensors in muscle or tendon via the posterior root ganglion of a spinal nerve; to the cord where it forms a synapse with an effector neurone, which transmits a motor response. More usually there is a connector neurone in the grey matter of the cord, which links the receptor and motor neurones by synapses in the posterior and anterior horns of the spinal grey matter.

Charles Sherrington (1857 – 1952) qualified in medicine at Cambridge and became professor of physiology at Liverpool University in 1895 and then at Oxford in 1913. Like Helmholtz he was concerned about the relationships between sensory inputs and motor responses and developed Hughlings Jackson's concept of increasing complexity from cord to higher centres. Sherrington realized that the basic unit of nervous response was the simple **spinal reflex** but that this could be modified by nervous feedback from higher centres in the brain.

In essence every **receptor or afferent neurone** is potentially able to communicate with **efferent or motor neurones** throughout the brain and spinal cord, limited only by transmission at the **synapses,** and modulated by the release of chemical neurotransmitters when a nerve impulse arrives. The body of a motor neurone in the anterior horn of the cord is a point of convergence of a number of efferent paths and the motor neurone is the **final common pathway** for active response.

The **cerebral cortex** initiates voluntary movements after integrating incoming signals and transmits impulses directly to the spinal cord via the **cortico-spinal or pyramidal tracts.** However the **cerebellum, basal ganglia and brain stem nuclei,** also receive and process signals from other brain regions, organs of special sense and pressure receptors (**proprioreceptors**) in muscles. These centres are able to coordinate and modulate muscle tone and regulate actions, which have been initiated by the cortex. **Voluntary motor action** is initiated in the **motor cortex** and impulses are transmitted via direct cortico-spinal pathways in the **pyramidal tracts** which decussate in the lower brain stem. **Involuntary motor actions**, concerned with posture, balance and integration of many different actions are modulated by centres in the basal gangia, brain stem and cerebellar cortex, which involve indirect, **extrapyramidal** pathways[7].

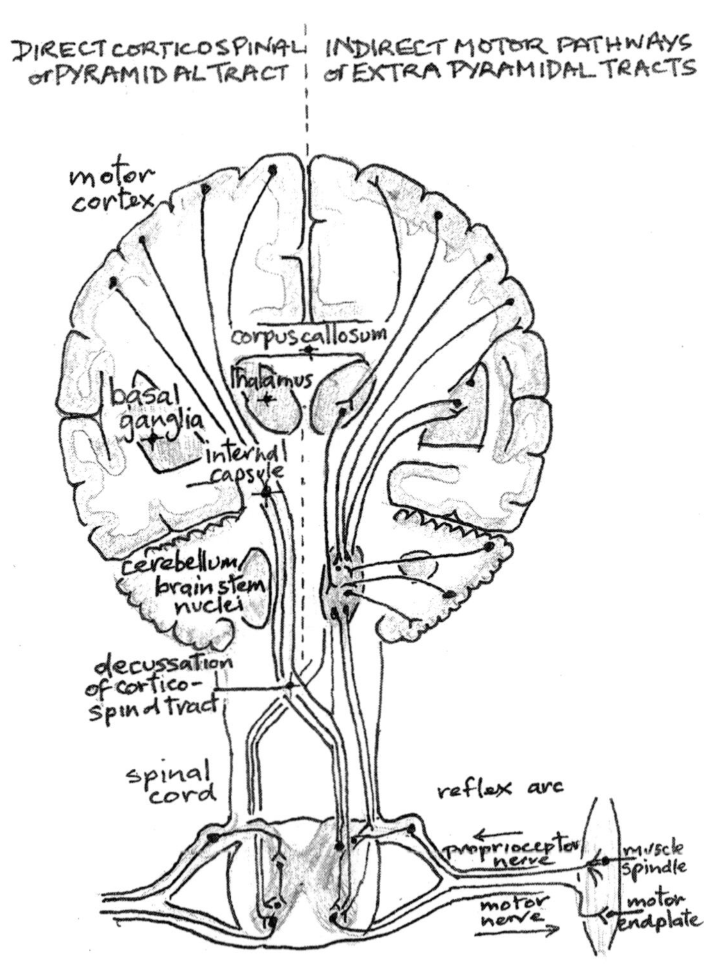

18. Diagram of the Brain and Spinal Cord showing **white and grey matter areas.** The **direct corticospinal (pyramidal) tract** on the left, is associated with **voluntary** movement.The **indirect pathways** shown on the right that involve basal ganglia, cerebellum and brain stem are **extrapyramidal** and unconsciously control and regulate body movements so maintaining **balance and posture.** The **pyramidal tracts** decussate (cross-over) at the pyramids in the lower brain stem.

Sherrington succeeded in bringing together most of the current understanding of nerve function and providing an account of nerve pathways and **synaptic transmission.**

He realized that the role of the central nervous system was to correlate all the sensory inputs and motor actions of an animal so as to give it an acute awareness of its external environment and the ability to respond appropriately, to threats or needs [7].

In 1904 Sherrington published *The Integrative Action of the Nervous System.* He was elected President of the Royal Society between 1920 and 1925. In addition to neurophysiology he made contributions to the understanding of cancer metabolism and infectious tropical diseases such as sleeping sickness, malaria, cholera and tetanus. He had a strong interest in the philosophy and science of the 16th and 17th centuries; he published *Man and His Nature* in 1938.

The neurosurgeon Dr Wilder Graves Penfield (1891 – 1976) was a Rhodes scholar who obtained his medical degree from John Hopkins University and studied neuropathology at Oxford with Sherrington. He continued Horsley's work on the surgical treatment of patients with severe epilepsy by ablating (destroying) nerve cells in the brain where seizures originated, and received support from David Rockerfeller to found a research institute. This was opposed by medical academics in New York so Penfield moved to Montreal in 1928 where he became Director of McGill University's Neurological Institute and Hospital.

Improvements in surgical technique and anaesthesia allowed Penfield to expose the brains of his patients whilst they were still conscious, using only local anaesthetic, and to observe their responses to electrical stimulation, so as to target epileptic foci more accurately and reduce the side-effects of ablation. The **neuro-electrode** technique also allowed him to map the sensory areas of the human brain, as well as the motor areas and to show their interconnections. In 1951 Penfield published *Epilepsy and the Functional Anatomy of the Human Brain,* which contributed to understanding the lateralizing of brain functions.

Today it is possible to map functional areas in the brain non-invasively using brain scanning techniques such as *Positron Emission Tomography (PET)* or *functional Magnetic Resonance (f MRI).* In 1967 Penfield was made a Companion of the Order of Canada.

In 1929 for the first time, amplifiers attached to electrodes on the scalp were used to detect minute undulations in electrical potentials arising from the brain, which were called **brain waves.** An entire record from 16 electrodes spaced evenly over the cranium is called an **electroencephalogram (EEG).** The intensities of the waves on the surface of the scalp range from 5 to 300 microvolts with frequencies of 0.5 to 50 per second. Different wave patterns are associated with varying states of consciousness. The most important use of EEG is in the assessment of patients with epilepsy and to identify sites within the brain responsible for focal fits which might be amenable to surgery.

There was popular interest in EEG and Albert Einstein, rightly regarded as a genius, was persuaded to have an encephalogram, to look at his brain waves whilst *'thinking about relativity'*. It was recognised that Einstein had an unusual brain and that although he was a genius, he didn't speak till he was three years old. Nor was he a success at school, which has given lots of people hope and consolation. Einstein had a sense of the ridiculous and obviously enjoyed the joke, but then became anxious about public interest in his brain. He insisted that when he died, all his remains, including his brain should be cremated. It is not known what his EEG showed.

After his death in April1955, Einstein's brain was photographed and preserved by the Princeton duty pathologist, a Dr Thomas Harvey, who had no special expertise in neuropathology or neuroanatomy. He thought Einstein's brain looked normal. It weighed 2.7 pounds, which is about average for a man in his seventies. After being sectioned into 200 blocks of tissue, it disappeared, following Dr Harvey into obscurity. In 1978 the remains were located in Wichita, Kansas, but escaped again.

The remains of Einstein's brain finally returned to Princeton in the early 1990s. Marian Diamond, a professor of neuroanatomy at the University of California studied sections through the general interpretive regions of the brain's cortex and compared the cell counts with several control brains. On the left side she found a significant increase in glial cells, which support the neurones metabolically, and suggested that this reflected a life time of heightened activity. It was not clear whether the increased ratio was either congenital or due to work hypertrophy over fifty active years of trying to reconcile quantum and relativity theories.

Then Sandra Witelson, a psychologist and neuroscientist at McMaster University, Ontario, looked at the remaining brain blocks and Harvey's original photographs. She saw at once that Einstein's brain was very different from nearly all the many others she had examined, which might explain his remarkable powers of insight. The pattern of the fissures indicated very marked enlargement of the inferior parietal lobe.

Witelson noted that although the appearance was rare, it was also evident in the brain of the great mathematician and geometer, Carl Friedrich Gauss (1777 – 1855) who had died a century before. She was accused of being a 'fissural phrenologist', because the interconnectedness of the brain occurs deep below the fissures, which only give approximate anatomical locations for the sites of different types of neurological function. However the inferior parietal lobe is known to be associated with mathematical and spatial reasoning and is close to the general interpretive region of the cortex.

The story of Einstein's brain is both shocking and farcical, but also full of pathos. The dissection of the preserved relics of a brain that contained a mind so perceptive, so generous and idealistic, also capable of human failings, seems idolatrous; exactly what he wanted to avoid. It raises the unfashionable concept of '*the ghost in the machine*'. In life Einstein's brain had been some machine; his theories and ideas haunt us still.

The marvellous ability of the human brain to integrate incoming sensations, memory, imagination and forward planning, with balanced and coordinated movements of the muscles of the trunk, limbs and eyes is demonstrated by the performance in ball games or by riding a bicycle [7]. Pressing bicycle pedals is initially a conscious action but then becomes semiautomatic like balancing on the bike and steering with the handlebars; eyes alert for potholes and traffic hazards; ears for noise of overtaking vehicles; nose smelling their exhaust fumes; wind on the face; at the same time able to make a shopping list, admire passing gardens or think about the weather.

The brain is the organ of the mind that enables us to think, to feel and have a sense of individual identity. Consciousness, mind and sense-of-self are not located in any one region of the brain but in a mysterious way, perhaps beyond the scope of physics, are consequences of the organization of cerebral function [8]. Some regard consciousness as being an *epiphenomenon* of brain function. This axiom may be descriptive but has no scientific explanatory power.

RESUMÉ

After 160 years the **electric** story came full circle; from Galvani's demonstration of bimetallic nerve stimulation and muscle contraction, which led to Volta's demonstration of direct current, then to Davy's use of Voltaic current to form ions in electrolytic cells and so to discover sodium, potassium, calcium and chloride These turned out to be crucial physiological ions for creating nerve membrane potentials, transmission of nerve impulses and muscle contraction. Awake and aware, the sensory network transmits information to the brain about the outside world, from all six senses. The brain and cord initiate many motor actions, modulated by reflex servo-mechanisms. When aware of external threats or opportunities, the brain configures a response and galvanizes the neuromuscular system into appropriate action. An integrated, bio-electric, neuro-muscular system is essential to the survival of a human being in uncertain external environments [7].

NOTES

1. **Coulombs Law** states that ***"Like charges repel and opposite charges attract with a force proportional to the product of the charges and inversely proportional to the square of the distance between them."*** The ***coulomb*** is the unit of electric charge and turns out to be equivalent to the charge on 6×10^{18}electrons.

2. The ***volt*** is the SI unit of ***electromotive force [emf]****; a difference of electric potential capable of sending a constant current of one* ***ampere*** *through a conducing circuit whose resistance is one* ***ohm.***

This is a version **Ohm's law** which defines the relationships between ***volts, amperes*** *and* ***ohms. "If the resistance of a conductor does not alter,the current that flows is proportional to the potential difference applied"*** or $V = IR$. (where V = voltage, I = current and R = resistance.)

The ***ampere*** is the unit of current and is equivalent to one ***coulomb*** of electric charge, flowing through a circuit **in one second.** The unit of electrical resistance is the ***ohm.***

Electric power is ***work done per second (joules/sec*** *or* ***watts)*** and is the product of ***potential difference (volts)*** and ***current (coulombs/second*** *or* ***amperes).***

3. In a zinc/copper electric battery the zinc plate, within the electrolyte attracts **negatively charged sulphate ions** (SO_4^-) the copper plate accumulates **positively charged copper ions** (Cu^+). The electron flow which results, drives the current, gives power to the circuit and comes from the chemical energy released by the transformation of copper sulphate to zinc sulphate. When this transformation is complete, the battery becomes *'flat'*.

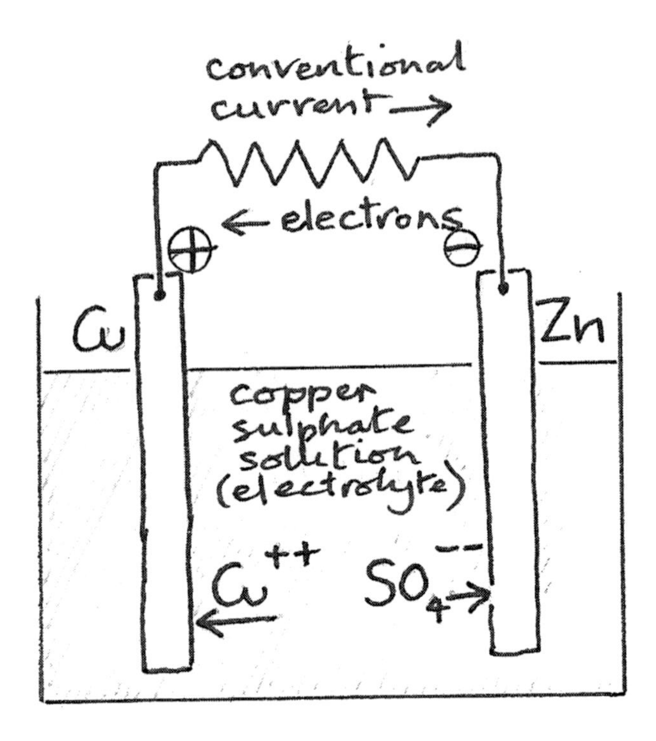

19. An Electric Battery consisting of copper and zinc plates immersed in copper sulphate solution..

An electric battery consisting of copper and zinc plates immersed in a solution of copper sulphate. When the plates are connected by a wire, a *conventional current* flows from the positive to the negative terminal. Copper ions accumulate on the copper plate and negatively charged sulphate ions on the zinc plate. Electrons flow in the wire from the negative terminal to the positive terminal in the opposite direction to the *conventional current,* or backwards.

4. Electrolysis of Water by passage of a direct current through immersed positive and negative electrodes was first attributed to Carlisle and Nicholson in 1800. **Hydrogen ions** (H^+) are attracted to the **cathode** and **hydroxyl ions** (OH^-) to the **anode. Hydrogen ions** unite to form **hydrogen molecules** (H_2). Two hydroxyl ions combine to form one molecule of water and release one **oxygen ion** (O^-); two of these then form an **oxygen molecule** (O_2); one for every two hydrogen molecules.

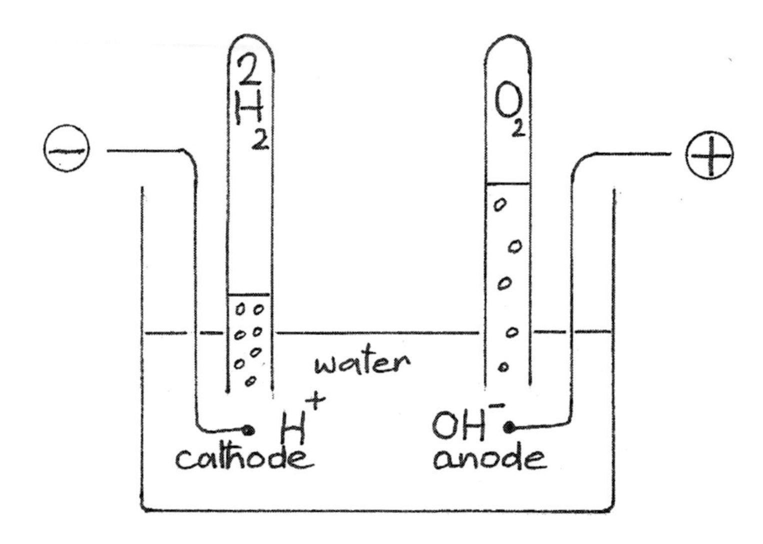

20. Hydrogen ions (H^+) are attracted to the **cathode** and **hydroxyl ions** (OH^-) to the **anode. Hydrogen ions** unite to form **hydrogen molecules** (H_2). Two hydroxyl ions combine to form one molecule of water and release one **oxygen ion** (O^-)**;** two of these then form an **oxygen molecule** (O_2)**;** one for every two hydrogen molecules.

5. A ***tesla*** (***T***) is that magnetic flux density, which will exert a force of one ***newton*** on a one metre wire carrying a current of one ***amp***. A ***tesla*** is about 2,000 times stronger than the earth's magnetic field.

6. It became clear that the **resting membrane potential of about –90 millivolts** within the nerve fibre is due to a high concentration of sodium ions (Na^+) outside the membrane induced by a **'pump'** in the membrane wall pushing sodium out and allowing some potassium ions (K^+) back in; in the ratio of 3 sodium ions *out* for 2 potassium ions *in* so creating the potential difference and **polarization** between the exterior (positive) and interior (negative) of the nerve axon. Before a depolarizing nerve impulse, this is the **resting potential** of the fibre.

Passage of an impulse, an action potential reverses polarity or depolarizes the membrane, so the membrane potential reverses from negative (-90 millivolts) to positive (+80 millivolts).

Depolarization is followed by a very brief **refractory period**, during which the membrane **'pump'** ejects sodium ions once again and restores the resting membrane potential. This process is **repolarization.**

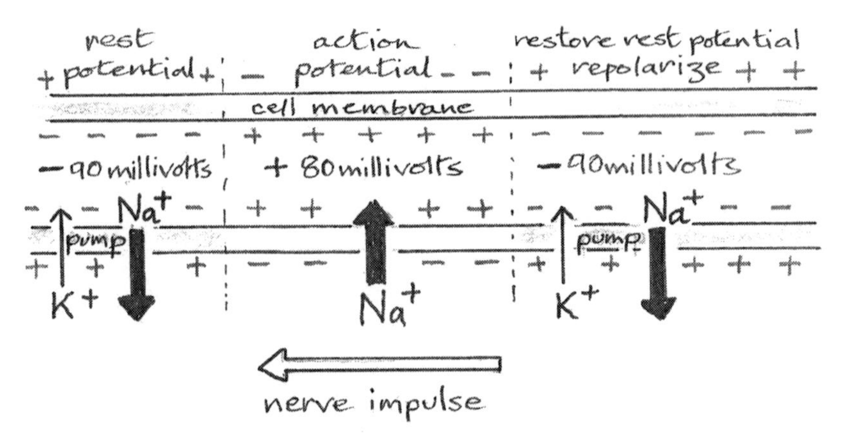

21. Movements of sodium ions (Na^+) and **potassium ions** (K^+) through the cell membrane of a nerve axon **at rest** and during the passage of an **action potential.** At rest sodium is ejected from the cell and a smaller amount of potassium enters so the internal **resting membrane potential is negative.** During **depolarization** sodium ions flood into the cell and reverse polarity so the interior **becomes positive.** During **repolarization** sodium ions are ejected once again.

7. Organisational management of the human neuromuscular (sensorimotor) system.

CEREBRAL CORTEX

SENSE ORGANS	SENSORY CORTEX	→	INTEGRATION	→	MOTOR CORTEX
			↕		
mouth	→	taste	awareness,	repertoire of movements,	
nose	→	smell	memory,	initiates voluntary action	
skin	→	touch, pain	imagination,	TRUNK	LIMBS
eyes	→	vision	planning,	global	skilled,precise
ears	→	hearing, balance	judgement,	movement,	movements,
		position	will-to-act	posture	fine actions
	↑		↕		↓
	↑		BASAL GANGLIA & THALAMUS		cortico-spinal
	↑		control and smooth regulation		or pyramidal
	↑		of muscle movements		tracts
	↑		↕		↓
PROPRIOCEPTORS		→	CEREBELLUM & BRAIN STEM		SPINAL CORD
pressure			control of rate, range, direction		
receptors in			and force of all movements.		↓ ↓
muscles and tendons			control of global movement, posture		MOTOR
respond					NEURONES
muscle action		←	MUSCLES	←	**final common pathway**
reflex circuits, tone			action, movement		reflex circuits, tone
			changed position		

Management of the Sensorimotor System. Sensory and motor circuits direct and manage musculoskeletal action with **'feed-back mechanisms'.** Motor impulses from the cortex come directly to motor neurones in the cord via the **pyramidal tract** or routed indirectly through the basal ganglia, thalamus, brain stem centres and cerebellum via the **extrapyramidal** tracts. The **motor neurone** is the **final common pathway.**

---

8. Professor Roger Penrose, the Rouse Ball professor of mathematics at Oxford University, has written extensively on artificial intelligence and the mathematical challenges of consciousness. In his book *Shadows of the Mind 1994,* he emphasized the non-computability of conscious thought and the need for a new physics to understand the mind.

VISIBLE LIGHT

*"Why Sir, we all **know** what light is;*
*but it is not easy to **tell** what it is."*
Samuel Johnson –' Boswell's Diary- April 1776'.

The scientific advances after the 17th Century were based on reliable observations that stimulated theorising and the proposition of natural laws. A reliable theory would explain all the initial observations and predict further findings. The invitation of science was to be watchful, to observe phenomena, hypothesize about their causation and then clarify theory with proof. The observer's eye, either naked, looking up a telescope or down a microscope had increased the objective understanding of the cosmos, material world and living things, including humans. *To see we need light.*

In physics, **light** is regarded as energy transmitted as electromagnetic radiation, which has properties of either waves or particles of energy (photons). Visible light covers a narrow band in the middle of the electromagnetic spectrum which extends from long radio waves to short x-rays and gamma rays with wavelengths of less than one millionth that of visible light.

Photon energy varies directly with frequency and inversely with the wavelength, so that x-rays and gamma rays have much greater energies, higher frequencies and shorter wavelengths than visible light or radio waves. Although they are invisible to humans, infrared and ultraviolet radiations are visible to some insects. Astronomers regard **all** electromagnetic radiations as **light** because although they have widely different energies and wavelengths, they travel at the same speed in space; 300,000 km per second.

Electromagnetic radiation is produced in four different ways, which characterize the SI units by which they are measured.

1. **Radio waves and microwaves** are produced by rapidly alternating currents in transmitter aerials or microwave coils and have wavelengths ranging from centimetres to hundreds of metres but are usually characterized by frequency, in Hertz.

   - one Hertz (Hz) is one cycle per second.

Microwave ovens use frequencies of about 3 Gigahertz (three thousand, million cycles per second). Radio waves are much less energetic with FM radio waves in the Megahertz (MHz) and AM radio waves in the kilohertz (kHz) ranges.

2. **Thermal radiation** includes all electromagnetic radiations produced by heating bodies to increasingly high temperatures; from a warm poker to the sun itself. It covers the range between infrared with wavelengths of micro-metres to ultraviolet with wavelengths of nanometres (1 nanometre = 10^{-9} metres – one thousand, millionth of a metre). Visible light is a narrow band in the thermal radiation spectrum with wavelengths of between 400-700 nanometres.

3. **X-rays** are made in an x-ray tube by a stream of high energy electrons, driven at very high voltage to bombard a metal target, such as tungsten. A standard diagnostic x-ray tube requires tens of thousands of volts, ranging from 40kV to 100kV, to produce radiation able to penetrate human tissues. For radiotherapy the voltages can be up to a million volts or more. The wavelengths of x-rays are extremely short, and measured in picometres (1 picometre = 10^{-12} metres - a million millionths of a metre), so x-rays are defined by their energies in kilo-electron volts (keV).

4. **Gamma rays** are emitted from the nuclei of radioactive atoms when they disintegrate and release energy. Gamma ray energies are comparable to high energy x-rays but are distinctive, with frequencies specific to each

radionuclide. Otherwise gamma rays are like x-rays and their energies are also measured in keV or MeV.

gamma rays / x-rays	ultraviolet	visible light infra-red	micro-waves	radio waves
I	I	I	I	I
10^{-14}	10^{-10}	10^{-6}	10^{-2}	10^{2}

wavelengths in metres

Thermal radiations, from infrared to ultraviolet show little or no penetration of human tissues, however at the extremes of the radiation spectrum, gamma rays, x-rays and radio waves are transmissible through the body and can be used to demonstrate internal anatomy using either radiographic, radio-nuclear or magnetic resonance imaging.

However this chapter will address thermally induced radiation, mainly sunlight and the visible spectrum in particular. Instead of '*To see we need light.*' it would be more precise to say '*To use our eyes we need light with wavelengths between 400 and 700 nanometres.*'

VISION

The interaction between the eyes and light, between the subjective experience of vision and the objective study of optics has always been and continues to be fascinating.

The ancient Greeks had complex notions about light and vision. Pythagorus (582 – 500BC) proposed a theory in which '*light*' consists of rays that act as 'feelers' that travel in straight lines from the eyes to the object. The sensation of '*sight*' is experienced by the eyes, when rays '*touch*' the object. Plato (427–347BC) believed similarly that rays of light originate in the eyes and travel in straight lines to the object being viewed, to produce vision. This concept underlined his parable, the '*Myth of the Cave*', which describes most of mankind as chained with its back to a fire in a deep cave. The flickering shadows on the cave wall are mistaken for reality. Plato regarded the education of a philosopher as a turning around of the '*eye of the soul*' so as to project light and vision beyond the mouth of the cave to reveal the true reality beyond.

Euclid (320–275BC) had a similar idea about rays of light spreading out from the eyes and concluded that the speed of light must be very great because when the eyes are closed and then opened, distant stars are seen instantly. Galen regarded the lenses of the eyes as the source of light rays.

The ancients had burning glasses and curved mirrors that focused sunlight to a point, indicating some knowledge of refraction and reflection. Ptolemy (100-170AD) measured the bending of a beam of light as it passed from air into water or glass due to refraction. Hero of Alexandria in the third century knew that when a narrow beam of light is reflected at a metal surface, the incident and reflected beams make equal angles with the surface.

The first person to study light, vision and the eye in a scientific manner was the Arabian scholar Alhazen (965-1039AD), also known as Ibn al-Haitham, who was from Basra. Initially he studied theology but then became intrigued by Aristotle's teachings and devoted himself to science and mathematics instead. He was invited to Egypt by al-Hakim, known as the Mad Caliph, to engineer a scheme to control the floodwaters of the Nile. Alhazen realized the impossibility of the scheme and feigned madness until after the Caliphs death.

During his time under virtual house arrest, Alhazen discovered the principle of the *camera obscura* in which bright light from outside a darkened room, entering through a small peephole at the front, projected an inverted image of the outside scene on the back wall of his room. He performed experiments on light and vision, and wrote on optics, geometry and astronomy. These experiments confirmed that light travels in straight lines. Alhazen also designed optical lenses for magnification and correctly proposed that the eyes passively receive light from objects, which contradicted the beliefs of Plato, Ptolemy and Galen.

Alhazen's book on optics *Kitab-al-Manadhirn* was published in 1038, translated into Latin by Witelo in 1270 and published by F Risner in 1572 with the title *Opticae Thesauris Alhazeni*. Witelo's translation was studied in many monasteries and monks became the first Europeans to use magnifying lenses to help them make illuminated religious manuscripts.

It was known by 1280 that elderly people with presbyopia could see better with magnifying glasses, which were popularised by the English monk Roger Bacon (1214-1294), also known as Dr Mirabilis, who followed the work of Alhazen. Short-sighted people had to wait until the 15th.Century for the invention of concave spectacle lenses to improve their vision. Bacon also experimented with the *camera obscura,* which was further improved by Geovanni Porta (1533-1615) who placed a lens in the enlarged aperture of a *camera,* which greatly increased the luminosity of the projected image. In his treatise *De Refractione* he endeavoured to explain the theory of lenses.

The first advance of theoretical significance since Alhazen was the experimental discovery of the *'Sine Law of Refraction'* in 1621 by Willebrorde Snell (1580-1626), professor of mathematics at Leiden. He gave lectures on refraction and the *Sine Law* but died before publishing the law himself. The law was published after Snell's death by Descartes[1].

OPTICS

In Europe by the end of the 16th Century, the idea that rays of light originated in the eyes had given way to the realization that light comes from the Sun, the stars and from the incandescent flames of fires and candles. The astronomer Johannes Kepler suggested in 1604 that the eye was like a *camera obscura* and that the retina, not the crystalline lens as previously supposed, was sensitive to light and responsible for visual sensation.

This concept was adopted by Descartes and in his treatise *Dioptrics,* published in 1637, he expounded his theory of vision, deduced from optical experiments with recently dead human and bulls' eyes. Placing an eye in the aperture of a *camera obscura* and replacing the outer coat of the back of the eye with a segment of eggshell, he described how light was refracted at the front of the eye and focused by the lens onto the retina to produce a tiny inverted image of the scene outside. Descartes postulated that this *'sensation image'* was transferred from the retina by the optic nerve, to the *'soul'* in the brain but he also recognised that we possess a more complex perceptual system than a simple camera would provide. He stated that the size, shape and position of objects are judged according to our knowledge or opinion and not necessarily in direct geometrical accordance with the image on the retina. Descartes regarded light as composed of rapidly moving globules and that colour came from the changing speed of the globules when refracted by water or glass.

As an observer and theorist Robert Hooke made important contributions to the understanding of light. He was aware that light does not always travel in straight lines but that there is faint illumination inside the shadow of an opaque body from a point source. This phenomenon was discovered by a Bolognese mathematician, Francesco Grimaldi (1618-1663) and became known as **diffraction.**

Most light travels in a straight line from the point source but a small proportion is diffracted sideways, out of the direct line.

To account for this Hooke proposed in *Micrographia* that the emission of light by a luminous body was due to rapid vibrations of small particles and that each particle propagates pulses that expand rapidly with equal speed in all directions, like circular ripples seen when a stone drops in water.

From this concept Hooke derived the ***Inverse Square Law*** that states– ***"The intensity of illumination at a distance from a light source is inversely proportional to the distance from the source."***

A practical problem affecting early microscopes and refractive telescopes was **chromatic aberration** of white light. Each colour component of the spectrum is refracted differently to form its own image and the different images do not coincide and vision is blurred. The problem increases with magnification. Boyle and Hooke had both observed iridescence in bubbles and thin glass plates. Hooke studied chromatic aberration and supposed that it was caused by the intrinsic nature of thin films, lenses and prisms, and so chose to use a reflective telescope for his astronomical work. He thought that red and blue were the two fundamental colours caused by *'confus'd pulses'* when light was refracted. For Hooke light was composed of pulses in the *ether* and colour was a disturbance of those pulses.

Huygens in The Hague and Newton at Woolsthorpe, became interested in the propagation, reflection and refraction of light after reading Hooke's discussions in *Micrographia* in 1665. In later years when they were both members of the Royal Society, Newton and Huygens had some collaboration, especially after 1688 and the Anglo-Dutch

Alliance. Huygens adopted and developed Hooke's undulatory theory and thought that pulses of light travelled through the all-pervading *ether* like waves on water or sound through the air. At each instant, every point on the crest of a wave can be regarded as the source of a secondary wavelet. The combination of wavelets propels the wave front forward and propagates the wave.

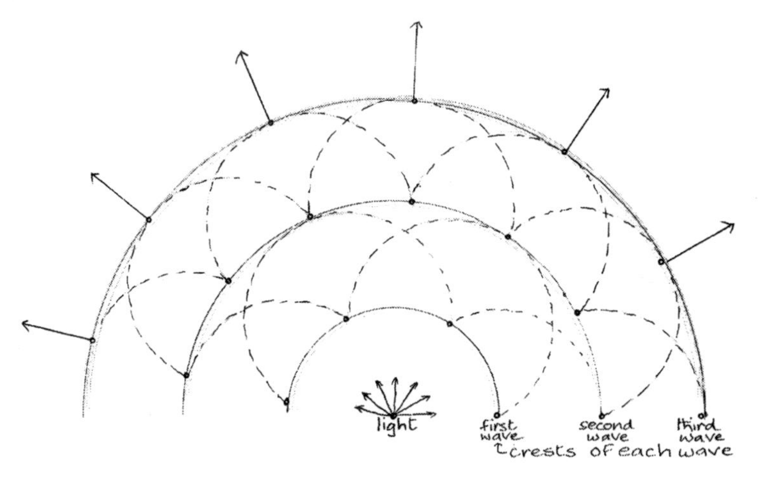

22. Huygens' Principle. The points on the crest of the first wave are the origins of wavelets which combine to form the second wave, then points on this are the origins of wavelets that form the third wave.

This concept became known as **Huygens' Principle**. His belief in it was based on the observation that two beams of light can cross without interfering with one another, as streams of particles might be expected to do; it also explained diffraction of light.

From air pump experiments Huygens knew that light could pass through a vacuum and assumed *ether* to be of very low density and high elasticity to accommodate the enormous speed of light that had been estimated by the Danish astronomer Olaus Roemer in 1675. Observing the time differences between the predicted and observed eclipses of Jupiter's moons at six month intervals Roemer found that light took nearly 1000 seconds to travel the diameter of the earth's orbit. The orbital diameter was known imprecisely in Roemer's time and so his estimate was less than the true speed of 300,000 kilometres per second.

Newton in *Opticks* pondered the difficulty of understanding the senses when these very senses were employed as the agents of understanding.

With this paradox in mind, he slid a bodkin into his eye socket and pressed the eyeball with its tip until he saw "*several white dark & coloured circles*".

He concluded "*To determine more absolutely what light is and by what modes or actions it produceth in our minds the Phantasms of* Colours *is not so easie.*"

Newton made his early light experiments in a darkened room with a hole in the widow shutter. A beam of white sunlight was focused onto a mounted prism and projected as a spectrum of colours onto the back wall. Newton described the series as violet, indigo, blue, green, yellow, orange and red.

When he passed a narrow band of the spectrum through a second prism he found no further colour change. From this and further experiments he concluded that white light is a mixture of colours, that a prism merely separates them and that the spectrum is an innate property of visible light. He kept this discovery to himself until 1672 when he informed the Royal Society that "*Light is a heterogeneous mixture of differently refrangible rays.*" Hooke's Theory of Colours was superseded within a few years.

Newton then studied the colours of thin plates and explained the concentric ring pattern, (*Newton's Rings*), by supposing that a ray of light alternates between *fits* of easy transmission and *fits* of easy reflection. He also proposed that the intervals between two *fits* of easy transmission to be dependent on the colour; greatest for red and least for violet.

Newton's preference was to regard light rays as "*multitudes of unimaginably small and swift corpuscles of various sizes and vigour.*" He was disinclined to accept Huygen's *undulatory or wave theory* because of the sharpness of shadows cast by opaque bodies. Grimaldi's diffraction he thought was due to distortion of the *ether* by a solid body.

Despite his uncertainty about wave theory, Newton calculated the intervals between the *fits* of easy transmission of the tiny corpuscles by measuring the distances between his concentric rings. The result of one forty-five thousandth of an inch (570 nanometres) is close to the wavelength of green light.

EYES AND BRAIN

Thomas Young (1773 – 1829), the highly intelligent son of a Quaker merchant banking family made many discoveries in physiological optics, physics and Egyptology, but is most famous for his attempts to win acceptance for a wave theory of light similar to that of Huygens. Whilst still very young he became a proficient Greek and Latin scholar and had read Newton's *Principia* and *Opticks*, Lavoisier's *Elements of Chemistry*, Black's *Lectures on Chemistry* and Boerhaave's *Methodus studii medici*. Between 1792 and 1799 he studied medicine in London, Edinburgh and then Göttingen where he obtained his medical doctorate.

In 1793 the mechanism of *visual accommodation* was unclear but respected authorities believed the eye adjusted its focus to different distances by changing either its length or the curvature of the cornea, where most optical refraction of light occurs. Young proved by experiments with ox eyes, focusing his own eyes under water and by examining patients with no lenses, that the ciliary muscle surrounding the lens adjusts its curvature for focusing. He showed that necessary changes in focal length could easily be achieved by changes to the shape of the crystalline lens, which has a higher refractive index than the transparent humours surrounding it. In 1800 he published his paper on the *Mechanisms of the Eye*.

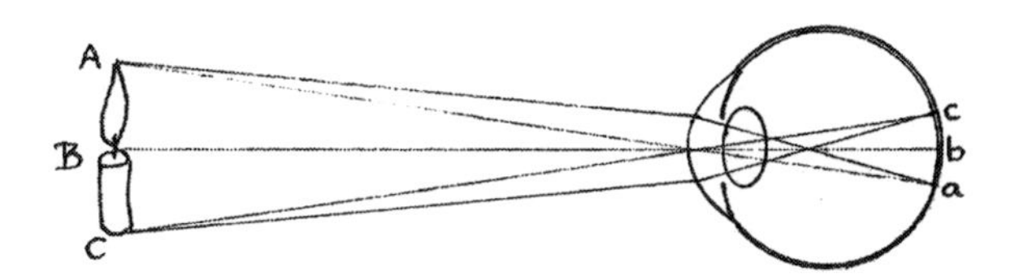

23. Light is focused by the cornea and lens onto the retina as an inverted image.

In its physical structure the eye resembles a spherical camera. At the front, the transparent cornea and lens focus light onto the sensitive retina at the back of the eye. Between the cornea and the lens is the aqueous humour and between the lens and the retina is the transparent vitreous humour each with a refractive index of 1.3; the index for the crystalline lens is 1.4. Most refraction and convergence of light therefore occurs between air and cornea with the modifiable lens fine-focusing light onto the retina. Young had shown that the lens is suspended by ligaments; and its shape is modified by the ciliary muscle. Immediately in front of the plane of the lens is the iris which like the diaphragm of a camera, controls the size of the aperture of the pupil and the amount of light falling on the retina.

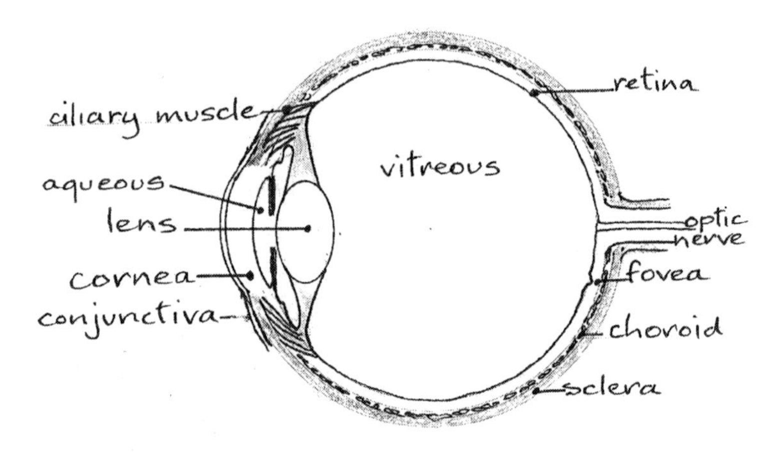

24. The eye

Except for the transparent cornea at the very front of the eye, the eyeball has three coats. Outside is the thick, opaque and fibrous **sclera**, white in colour, which maintains the shape of the eye and to which the external ocular muscles are attached. Then a vascular layer called the **choroid** whose blood vessels supply nutrition to internal eye structures and which is dark brown to prevent light reaching the retina, except through the pupil. The innermost delicate membrane is the light sensitive **retina**, pink in colour, which is in contact with the transparent membrane containing the vitreous humour.

Kepler had discovered in the 17th century that the retinal image of external objects is inverted, as in a *camera obscura*, so that light from above is projected below and from objects on one side to the opposite side of the retina. This occurs in both eyes, so that light from one side is projected on the opposite sides of each retina. The tracts in the optic nerves and brain are so arranged that nerve impulses from the lateral (outside, nearest the temple) sides of the retinas go to the visual cortex on the same side and impulses from the medial (inside, nearest the nose) sides of the retinas cross over at the optic chiasma (crossing) to reach the opposite visual cortex. The result is that visual information from objects on the right side is processed in the left visual cortex and vice-versa.

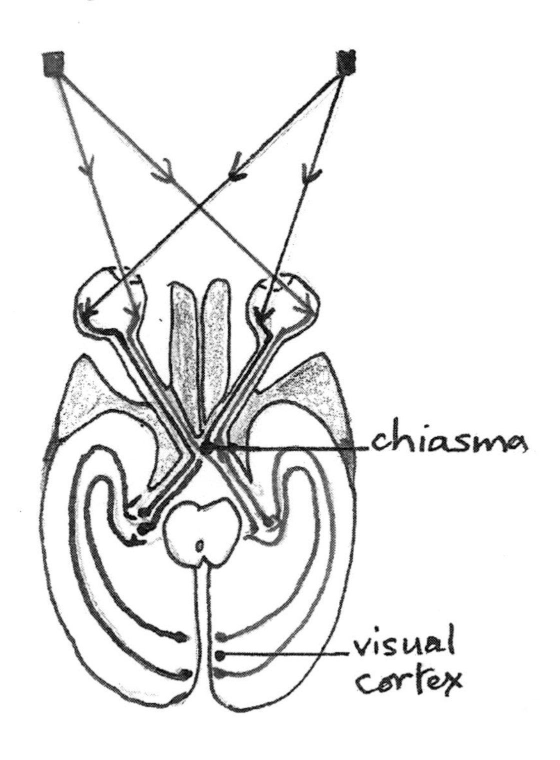

25. The Visual Pathways. Visual information from the right field is projected on the left side of each retina and processed by the left visual cortex. Conversely on the opposite side.

Young suggested in 1801, that the retina responded to just three different colours and stated that "*The perfect sensations of yellow and blue are produced respectively, by mixtures of red and green and of green and violet light, and there is reason to suspect that those sensations are always compounded of the separate sensations combined.*"

Young arrived at this conclusion by reasoning that the retina could not have separate receptors for each of the 200 separable hues because acuity of vision is almost as good in coloured as in white light. More importantly Young performed experiments with beams of light of the three different colours projected on a screen, and showed that any spectral hue could be produced by adjusting the relative intensities of blue, red and green light beams. He could make white, but not black or non-spectral colours such as brown. Young had shown that the eye effectively mixes the three colours to which it is sensitive to give the whole range of hues that we perceive.

Clerk Maxwell later wrote *"It seems a truism to say that colour is a sensation; yet Young by recognising this elementary truth, established the first consistent theory of colour.... and sought for the explanation not in the nature of light but in the constitution of man."*

Although much more is now known about the physical and photochemical mechanisms of colour vision, Young's ***three colour theory*** still holds up. His optometric methods were the foundation of ophthalmic optics.

In 1801Young was appointed professor of natural philosophy, editor of journals and superintendent of the newly established Royal Institution; he also started his London medical practice and continued his research. His lectures were too erudite for his students, his patients were demanding and the managers of the new institute displeased. Like the intelligent man he was, he resigned the professorship, transferred his practice to the seaside resort of Worthing and relinquished direction of the Royal Institution to Humphry Davy who proved more popular.

Young continued his research in London and at Cambridge where he also obtained a British medical degree in 1803.

WAVES OR PARTICLES

So far Young had been mainly concerned with visual optics and physiology. He now transferred his attention to the nature of light. The foundations for theories of light had been laid by Newton and Huygens but whether the corpuscular or undulatory theory best fitted the facts was still controversial. The corpuscular theory was dominant because it was simpler mathematically and was thought to have been favoured by Newton, however Young knew from *Opticks* that Newton was not adamant; indeed his theory of '*easy fits*' had some oscillatory features and Newton had remained on good terms with Huygens who he respected as a scientist and mathematician. Young became a new champion for wave theory by stating that '*Some considerations may be brought forwards, which tend to diminish the weight of objections to a theory similar to the Huygenian.*'

His argument was that '*All impressions are known to be transmitted through an elastic fluid with the same velocity*' and was probably derived from Newton. Young argued that the constant speed of light indicated that this was due to impressions or waves in the *ether*, similar to the vibrations of sound in air, however to transmit light waves so fast, meant that the *ether* was highly elastic.

In May 1801 Young discovered his ***Principle of Interference*** that stated ***"When two undulations (waves) from different origins coincide either perfectly or very nearly in Direction, their joint effect is a Combination of Motions belonging to each."***

Perhaps the Principle occurred to him as he watched waves on the beach at Worthing. Young then performed a crucial experiment in which light from a point source passed through two thin slits in an opaque flat screen and projected an image of many alternate bright and dark bands onto another screen. The bright bands showed chromatic aberration, in other words each of the bright bands was composed of smaller bands representing each colour of the spectrum.

According to *corpuscular theory* only two bright white bands should be projected onto the second screen whereas *wave theory* could explain the alternate bands by interference of light waves. The light passing through each slit would travel very slightly different distances to the second screen, of the order of 400 to 600 nanometres, and at different angles, making wave interference possible. There would be summation of light intensity when the waves were in phase and loss of intensity when waves were out of phase.

The similarity of his spectral bands to Newton's rings meant that they too must be due to the same phenomenon of interference. Young also realized that Newton had been mistaken in thinking that light travels faster in a dense medium than in air. In fact it travels more slowly and this would account for the refraction of light according to Snell's Law. He concluded that light must be an undulation in the *luminiferous ether*. Young became a strong advocate of the *Wave Theory of Light* [2].

Young's experiments and explanations were essentially qualitative and he was not able to demonstrate mathematical proofs, so that the immense importance of his concepts was not realized initially.

As Galileo and his colleagues had to overcome the opposition of Aristotelian scholarship, so Young and his French co-conspirator, Augustin Fresnel had to overcome the prestige of Newton; easier in France than England. They had shown that wave theory could explain Newton's rings satisfactorily as due to interference between periodic trains of light waves reflected from two surfaces nearly in contact. An interference pattern can be seen by looking at a bright light through a small white feather.

Young's London practice recovered and he was elected to the Royal College of Physicians in 1808, then physician to St George's Hospital. In 1813 he started to translate the demotic script of the Rosetta Stone and established its relationship to the Egyptian hieroglyphs. He wrote the article on *'Egypt'* for *Encyclopaedia Britannica,* was on the Board of Longitude and of the *Nautical Almanac.* His final work was *the Encorial Egyptian Dictionary,* published posthumously in 1830.

The Parisian, Augustin Jean Fresnel (1788-1827) was an engineer and physicist who supported the *undulatory theory,* even more strongly when he heard of Young's research and the phenomenon of interference. Fresnel was able to give mathematical substance to explanations of diffusion, reflection and refraction based on the wave theory, which ultimately led to its acceptance. Huygens had already shown how the diffusion of light around the edges of an opaque object could be explained by his principle of secondary wavelets. Reflection was explicable by both corpuscular and wave theories. Refraction was the challenge for corpuscular theory. Although Snell's law predicted the manner in which light was refracted on entering or leaving a denser medium it did not explain the phenomenon. Fresnel was able to give a convincing mathematical explanation of *refraction* using *wave theory* [3].

Because white light is composed of light of different frequencies, *changes of wavelength* and *speed* from *air* into *a transparent medium* such as glass are also different. The refraction of light of different colours varies slightly, even in the same medium. Consequently the angle of refraction for each colour is different. Red light has a lower frequency and longer wavelength than blue light; its speed is reduced less than blue light and so it is refracted less than blue light, on entering a medium like glass, obliquely. Blue light is slowed more and refracted at a greater angle than red but not so much as violet. This refractive separation of colours accounts for the phenomenon of Newton's spectrum. The ability to analyse the visible spectrum created the new science of **spectroscopy**, which provided a means for measuring the wavelengths of different colours of light.

Iceland spar, composed of calcium carbonate crystals, was known by Huygens to have unusual optical properties and Young suspected in 1817 that light may have *"some property resembling polarity."* Before this he had assumed that light waves were longitudinal like sea or sound waves. The idea that they might be transverse, like those of a vibrating string, and could be polarized in the transverse plane occurred to Young. He mentioned it to Fresnel who became totally convinced. Fresnel applied the concept of transverse undulation successfully to a detailed mathematical analysis of known optical phenomena: reflection, diffraction, partial reflection, refraction and polarization, which subsequently established the wave theory of light.

Fresnel and his brother were popular in England for the lenses that they developed for light houses and which saved many lives by improving coastal navigation. Fresnel was elected to the Academie des Sciences and the Royal Society.

THE MICROSCOPE

In 1610 Galileo made a compound microscope with two magnifying lenses, one at the eye-piece and the other near the object of study. Unlike his telescope it provided him with no significant discoveries. There was little improvement for over a century on the instruments devised by Hooke and Leeuwenhoek because of aberration of light by lenses which caused image blurring that became worse with magnification.

Chromatic aberration was due to spectral separation of light of different colours which focused at different planes. **Spherical aberration** was caused by light passing through the peripheral part of the lens focusing differently from light passing near the centre. This blurred the view even after achromatic lenses were invented. But in 1830 the physicist and microscopist Joseph Jackson Lister, father of the surgeon Joseph (later Lord) Lister, designed an achromatic objective lens which corrected for spherical aberration. The improved resolving power of Lister's instrument transformed microscopy. Clear images of fine slices of plant and animal tissues revealed their cellular structure in greater detail than ever before.

Medical pathologists used the new microscopes increasingly to examine diseased tissues obtained at autopsy and in 1858 Rudolf Ludwig Virschow (1821 – 1902), published his encyclopaedic *Cellular Pathology* which identified the tissue cells as the seat of disease processes. The classical *Theory of Humours* gave way to **cellular pathology** as revealed by the microscope. In the same year Louis Pasteur (1822 – 95), the professor of chemistry at Lille, published his first important paper on fermentation and demonstrated that the ferments were micro-organisms. Pasteur studied the fermentation of milk, wine and beer, then infectious diseases such as cholera, anthrax and rabies for which he developed inoculations. Pasteur had established the science of **bacteriology.** Robert Koch (1843 – 1910), a physician from Berlin, studied bacteria microscopically using special staining techniques and identified the organisms responsible for cholera, anthrax and tuberculosis. By 1900 more than twenty important pathogenic bacteria had been identified microscopically and several had been shown experimentally to be contagious.

Joseph Lister (1827 – 1912) qualified in medicine and surgery in London in 1852. Using one of his father's microscopes he explored the fine muscles of the iris of the eye and described how they dilated and contracted the pupil to control the amount of light entering the eye at different levels of illumination. He then studied inflamed tissues and the microscopic changes in small vessels surrounding infected wounds.

Lister was appointed professor of surgery in Glasgow in 1859. By 1865 he realized that nearly half his patients who had survived limb amputations went on to die from sepsis and gangrene. Lister became an early disciple of Pasteur and recognized that fermentation and putrefaction were similar phenomena; that both were caused by infective micro-organisms. He began to dress his patients' wounds with an antiseptic solution of carbolic acid and to operate under a carbolic spray. Within two years he reduced the post-operative mortality rate of his patients from 46 to 15 percent. Initially there was strong opposition to his *'germ theory'* but eventually the effectiveness of his *'carbolic treatment'* became generally accepted and widely adopted. Lister became a Peer in 1897 and a Member of the Order of Merit in 1902. His father's microscope had played a significant role in establishing the sciences of microscopic anatomy (histology), cellular pathology (histopathology) and bacteriology (microbiology).

VISUAL PERCEPTION

Microscopic studies of the retina by Trevanus in 1835 revealed a layer of rod and cone shaped cells connected by fibres to neurones with axons in the optic nerve. He assumed correctly that these cells were light sensitive but made the natural mistake of thinking that this layer was on the inside nearest to the transparent vitreous humour with neurones and capillaries outside. Heinrich Müller (1820 – 1864), the professor of anatomy in Wurzberg looked more carefully and found that the human retina was 'inside-out' and that the rods and cones were on the outside so that light had to pass from the vitreous humour, through layers of nerve fibres, nerve cells and blood vessels before reaching them; except at the back of the eye where in most people there is a small yellow spot called the *macula*. This measures only one millimetre in diameter and has a small pit in the middle called the fovea where there are no overlying fibres or blood vessels between the vitreous and the receptor cones.

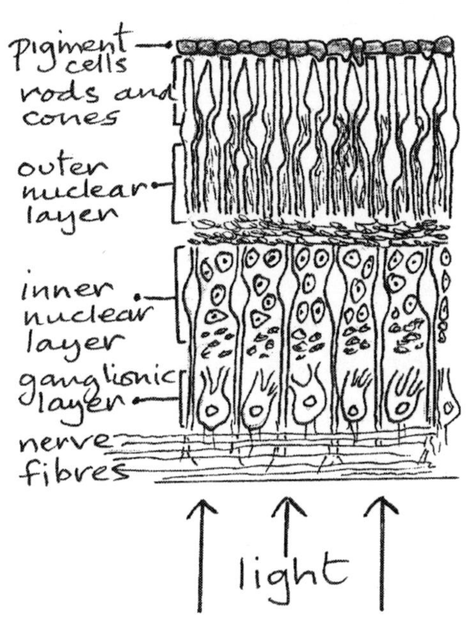

26. The layers of the retina.

In 1856 Müller published his book *Anatomical and Physiological Studies of the Retina of Humans and Apes.* His research revealed that **cones** function in daylight conditions and give colour vision (**photopic vision**) whereas **rods** function best in nocturnal conditions, at low illumination and give only shades of grey (**scotopic vision**). The distribution of rods and cones varies in different parts of the retina. The fovea, the region of most acute vision in daylight contains only cones packed closely together. The small size of the fovea corresponds to less than half a degree of arc so that the area of detailed central vision forms a small though important part of the whole visual field. As it contains no rods central vision is very poor in the dark so scotopic perception of detail is best just off-centre of the visual field.

Outside the fovea there is a mixture of rods and cones with rods predominating towards the periphery of the retina. As a consequence there is high definition of colour and spatial relationships in the small central field of *foveal vision,* which disperse towards the periphery, matching the retinal distribution of rods and cones. In a small area a few millimetres away from the macula, called the optic disc, there are no rods or cones. Here nerve fibres from the retina converge to form the optic nerve and the disc causes a '*blind-spot*' in the visual field.

In the University of Köningsberg, before he moved to Berlin, von Helmholtz was advancing optical physiology with studies on the visual perception of space, of movement and colour. In 1851 he invented the *ophthalmoscope* for examining the inside of the living eye; especially the retina. With modifications this instrument is still in use today in medical clinics around the world. An understanding of the functional anatomy of the eye and brain, and visual mechanisms of perception began to emerge. Helmholtz believed that visual perception consists in the formation; testing and correction of visual hypotheses based on empirical feedback and showed that the eye is controlled by principles of probability calculus such as the method of least squares. He regarded the eye as being a mathematician.

The eyes and brain use several strategies in perceiving the *position* of objects in surrounding space. For example, in estimating the distance of a familiar object from the eye, the size of its image projected on the retina will give an initial clue, then focusing the lens will confirm the correct distance. Binocular vision involves coordinated movements of both eyes to centre on objects to give stereoscopic perception of their spatial relationships to the observer and to one another. Even with a single eye, head movement and parallax can be used to gauge distances because near objects will appear to shift more than distant ones. The *visual feedback circuits* involving the retina, optic pathways, cortex and oculomotor centres respond with extreme rapidity and iterate to give successively closer approximations of spatial relationships until a coherent perception is achieved. Sometimes, as with ambiguous figures like the Necker cube, two possible, but mutually exclusive, spatial perceptions alternate.

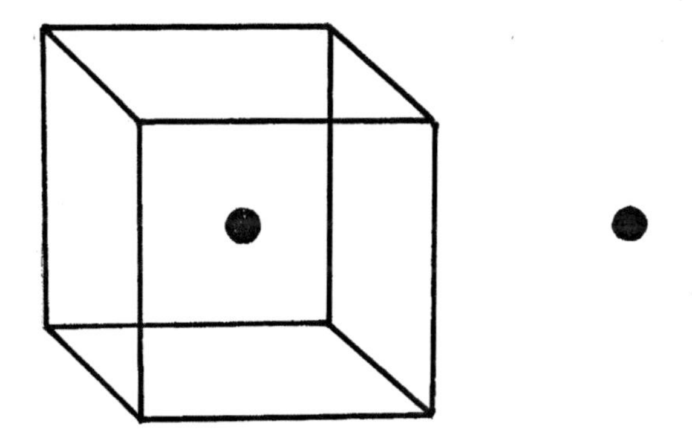

27. The Necker cube. Whilst looking at the cube with both eyes, two spatial perceptions alternate. With left eye closed, focus on the black dot inside the cube with the right eye. At a distance of about 20 cms the black dot on the right will disappear in the 'blind spot'.

Perception of *movement* of objects results initially from an image moving across the retina stimulating different areas, and for unexpected movements often in the periphery of the retina. Very quickly the eye and brain go into 'tracking' mode, coordinating ocular movements so that the moving image is captured and held in central foveal vision.

Perception of *colour* is complex and involves elaborate photochemical mechanisms. Helmholtz repeated Young's experiments and was able to measure the wavelengths of visible light more accurately and confirmed the *three-colour hypothesis*. This is often called the Young-Helmholtz theory, after these two very different physicist-physicians; Young, a gentleman scientist and London society doctor; Helmholtz, the Director of the Institute of Physics in Berlin and ex-army surgeon.

At the same time, James Clerk Maxwell (1831 – 1879), a young Scottish mathematician and physicist was independently studying the three-receptor theory with greater precision and published his first paper in 1852, possibly before Helmholtz. Maxwell demonstrated that colour blindness was due to ineffectiveness of one or more types of receptor. He was the first to project a colour photograph on a screen, in front of an audience that included Faraday at the Royal Institution in 1861.

Maxwell was born in Edinburgh and at the age of only fourteen had rediscovered Descartes' mathematical proof for the description of an ellipse. After three years at Edinburgh University he moved to Cambridge where he became second wrangler in mathematics and a fellow of Trinity College. Whilst still a student in Edinburgh, Maxwell began experiments on colour mixing by observing hues generated by coloured sectors on a rapidly spinning disc and was able to formulate quantitative colour equations employing red, green and blue as primaries. He distinguished three variables – of hue (spectral colour), tint (degree of saturation), shade (intensity of illumination) and established the science of *colorimetry*. He plotted the colour response curves of many different observers, including some with colour blindness, to precise wavelengths of light.

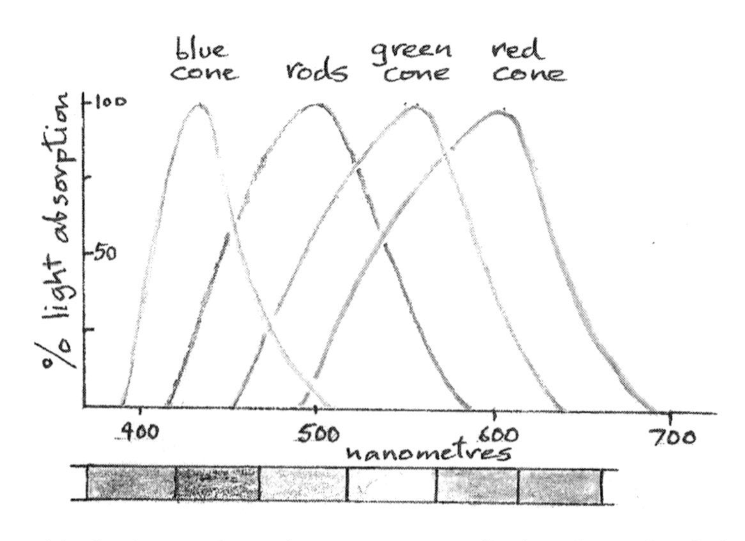

28. The fundamental visual response curves of rods and cones in relation to wavelength of light.

The *photopigments* in the receptors are four related chemicals that are 'bleached' by very small quantities of light energy (*photons*) with the release of electrons that initiate a nerve impulse. In the rods the pigment is *rhodopsin* with a peak sensitivity of 500 nanometres, which though in the green region of the spectrum only gives perception of shades of grey. The pigments in the three types of cones, called *photopsins* are chemically similar to *rhodopsin* but with minor modifications affecting their peak sensitivities. *Blue sensitive pigment* has a peak at 430 nanometres, *green* at 535 nanometres and *red* at 585 nanometres. Rods and cones are only a few microns in length and not much longer than the light waves to which they are sensitive. Once *'bleached'* the pigments are restored by vitamin A related compounds from the adjacent pigment cells in the outermost layer of the retina.

As a student in Edinburgh, Maxwell had been involved with research in spectroscopy and colour photography, and he was very well aware that there were invisible radiations in sunlight, above and below the visible spectrum from the previous work of Herschel and Ritter. Friedrich Wilhelm Herschel, later Sir Frederick William Herschel FRS (1738 – 1822), was one of the most talented 18th century astronomers who discovered *infrared radiation* by passing sunlight through a prism and holding a thermometer just beyond the red end of the visible spectrum. The thermometer showed an increase in temperature, from which Herschel concluded that there must be an invisible form of light which he called *'heat rays'*. It was later shown that radiant heat could be reflected, refracted and polarized in the same way as visible light. Beyond the other end of the visible spectrum, the German chemist and physicist, Johann Wilhelm Ritter(1776 – 1810) had used chemical changes in silver chloride to identify invisible *ultraviolet* light which he called *'chemical rays'*.

ELECTROMAGNETIC WAVES

Soon after his Cambridge graduation in 1854 Maxwell began his researches on ***electricity and magnetism***. The influences on him were those of Faraday and William Thomson (later Lord Kelvin). Faraday's great discoveries were electromagnetic induction, laws of electrical and magnetic fields and the laws of electrochemistry. Faraday had also demonstrated the phenomenon of magneto-optical rotation. With a powerful magnet he had rotated the plane of a beam of polarized light.

At the same time, Antoine-Cesar Becquerel (1788 – 1878), professor of physics in Paris and grandfather of Henri, had shown a *photoelectric effect* by shining light onto the plates of a Voltaic cell which altered its electromotive force. Faraday's and Bequerel's experiments suggested that light, magnetism and electricity might be linked.

In 1860 Maxwell became professor of natural philosophy at King's College, London and was elected a Fellow of the Royal Society in 1861. Shortly after, he came to the conclusion that an electric field gives rise to displacement of electricity in the *ether* proportional to the strength of the field and a corresponding change occurs with the magnetic field. From this concept he calculated that the speed of propagation of electrical and magnetic fields in the *ether* and found that it was very similar to the known speed of light.

In 1862 he wrote, "*We can scarcely avoid the conclusion that light consists in transverse undulations of the same medium which is the cause of electric and magnetic phenomena.*"[iii]

He concluded that light waves had a dual form, consisting of waves of magnetic force and electric displacement with motions perpendicular to each other.

Maxwell formulated the four ***General Equations of the Electromagnetic Field*** in a fully developed form in *Electricity and Magnetism* in 1873. His equations gave formulae for refraction, reflection and scattering of light consistent with each other and with experiment. He had shown how the fundamental relationship between electricity and magnetism, revealed by Faraday's experiments, provided a theory of complete generality in which energy is transmitted by electromagnetic waves.

Maxwell's theory received impressive experimental confirmation within ten years of his death and became widely accepted. However it depended on the concept of *luminiferous ether* as the medium for transmitting light waves. Unlike Huygens' and Young's elastic *ether*, Maxwell's was conceived as a perfect electric insulator in which electromagnetic waves could be propagated and transmitted.

The other factor of significance was that the speed of light could be derived from intrinsic electromagnetic properties. Since Roemer in 1675, *the speed of light* had been known approximately but had no deeper significance. After Maxwell its role seemed more fundamental.

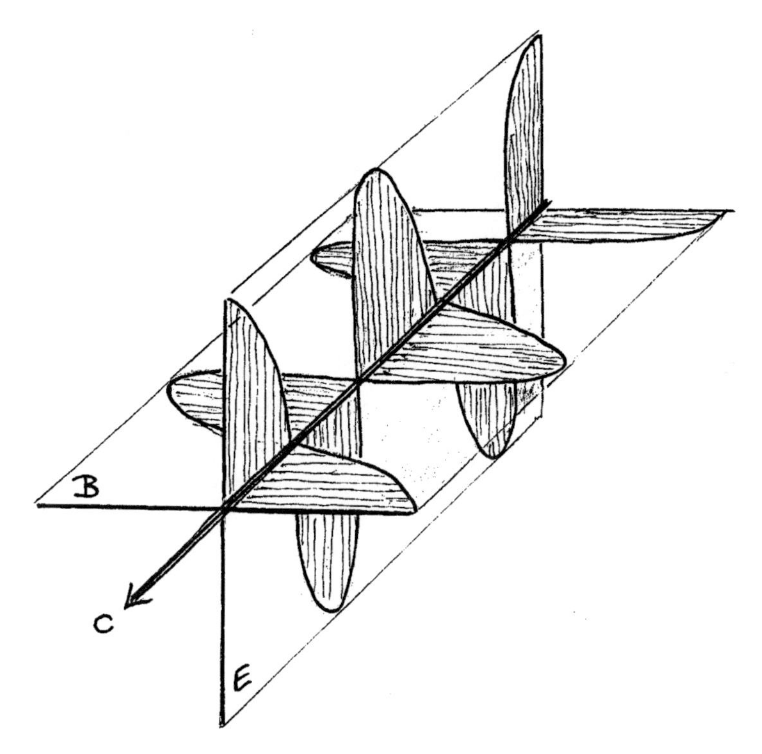

29. Plane-polarized electromagnetic wave. The electric field varies in the vertical plane **E** and the magnetic field varies in the horizontal plane **B**, for a wave propagated in direction **c**.

In 1874 Maxwell became the Cavendish professor at the University of Cambridge where he established the Cavendish Laboratory in Free School Lane. It was in this laboratory that JJ Thomson, his successor was to discover the electron in 1897. In 1879 Maxwell developed gastric cancer, but continued his teaching and research for as long as he could before his death at the age of forty-eight.

James Clerk Maxwell was the most outstanding British mathematical physicist since Isaac Newton. He extended Faraday's theories of electricity and magnetic lines of force to understand the nature of *light* as *electromagnetic radiation,* formulated the *kinetic theory of gases,* which was the foundation of *statistical physics,* confirmed the *three-colour theory* of vision on an exact numerical basis and proved that the *rings of Saturn* were composed of small solid particles.

AFTER MAXWELL

By 1880 the major known compartments of physics had united; the energies of *force, sound, heat, light, electro-magnetism* and *chemical action* had been found to be interchangeable. Only *gravity* was exempt. It appeared to some that the great challenges of physics had all been addressed.

Maxwell's theory of electro magnetic radiation embodied in his four General Equations was not followed up in Britain. A physics professor in Karlsruhe, Heinrich Hertz (1857 – 94), who had been one of the brightest students of Helmholtz at the Institute of Physics in Berlin, was attracted to Maxwell's field theory and proceeded to verify it experimentally. Hertz designed an electrical circuit to generate electromagnetic energy as invisible waves, with a rapidly alternating frequency of 100,000 cycles per second (100 kHz) and with a wavelength of 3 kilometres. Hertz then devised a detecting circuit tuned to resonate at the same frequency and was able to transmit waves the length of his laboratory. Guglielmo Marconi showed how *Hertzian waves* could be transmitted over longer distances and sent the first radio signal across the Atlantic in 1901.

The Polish American physicist, Albert Abraham Michelson (1853 – 1931) came to the United States at the age of two. It is alleged that as an impoverished young man he loitered at the door of the White House and persuaded the President, who took a daily walk in public, to secure him a free place at the US Naval Academy where Michelson studied physics. After further study in Europe he became professor of physics in Ohio, then Massachusetts and finally, chief professor at the Ryerson Physical Laboratory in Chicago.

This determined individual had supreme skill at designing optical equipment of incredible accuracy, in particular an interferometer, for measuring the speed of light. Michelson was assisted in his research by his student Edward Morley. In 1881 the *Michelson-Morley Experiment* was performed which found that the speed of light from the sun was identical, either in line with earth's direction towards or away from the sun, and that there was *no* evidence of *ethereal drift,* as would have been predicted by Maxwell's electromagnetic wave theory. At the time, not too much was made of the inconvenient discovery that in space the *speed of light was constant.*

In 1894, Max Planck, who was professor of physics at Berlin University, was commissioned by an electricity company to find ways of increasing the efficiency of electric light bulbs; the most light for the least energy. Planck's major interest was in thermodynamics and the physics of heat transfer, which he acquired when he had been a student and research associate of von Helmholtz in Berlin. He carried out his commission by studying what British physicists choose to call '*black body radiation*' and what German physicists call '*cavity radiation*'. Perfect '*black bodies*' are those which are hypothetically perfect absorbers and perfect emitters of radiation. The question was – "*How does the intensity of the electromagnetic radiation emitted by a 'black body' depend on the frequency of the radiation (colour) and the temperature of the body?*" The '*black body*' in this instance being the incandescent filament of a light bulb.

In 1900 Planck presented his ***Law of Black Body Radiation,*** which contained the supposition that electromagnetic radiation could only be emitted as *quanta of energy* [4]. Planck accepted the notion of *quanta* or small packets of energy, with extreme reluctance because he had been compelled to use Boltzmann's statistical interpretation of the Second Law of Thermodynamics. Planck called it "*an act of despair....was ready to sacrifice any of my previous convictions about physics.*" However his belief in the power of logical thought, based on experimental facts was so strong that he had proposed a law contrary to his previous convictions, because no other resort was possible.

Planck considered *quantization* to be just a mathematical stratagem, however his attempts to reintegrate *action quanta* back into classical physics were unsuccessful. Initially he assumed that light *quanta* were manifestations of *Hertzian oscillators (atoms)* in the heated body and not intrinsic to light itself. However Planck's recognition that the energy states of a physical system could be discrete was the birth of quantum physics and he was awarded the Nobel Prize for Physics in 1918.

After the First World War, Planck became the highest authority in German physics and worked hard to re-establish research facilities and positions, despite personal misfortune. His first wife had died of tuberculosis in 1909, his eldest son had been killed at Verdun and both his twin daughters had died in childbirth. After 1933 when the Nazis seized power he was distressed to see so many Jewish friends and colleagues expelled from their posts. In 1936 he was forced to resign as president of the Kaiser Wilhelm Institute, for teaching *'Jewish science'*. Planck's home in Berlin was completely destroyed in an air raid in 1944, then his son Erwin was executed for his participation in the failed attempt to assassinate Hitler. In 1948, the Kaiser Wilhelm Institute was renamed the Max Planck Institute.

In 1899, Philipp Lenard (1862 – 1947) the professor of physics in Heidelberg found that light shining onto a metal surface in an evacuated tube produced cathode rays, which JJ Thomson in Cambridge had shown to be electrons. Lenard studied this *photoelectric effect* and observed that the brighter the light, the more electrons were displaced from the metal but at the same velocity. Using light of different colours, he showed that electrons moved faster with light of higher frequency; much faster for ultraviolet light than blue and for blue faster than red light.

At the beginning of the twentieth century, *physics*, which after Maxwell had been considered to have resolved most of its major controversies, had suddenly become incomprehensible and complicated once again.

The notions of minute *quanta of energy*, the *constant speed of light* and the *photoelectric effect* defied explanation by the laws of classical physics. *Physics* needed a genius clarify these problems.

EINSTEIN

Albert Einstein (1879 – 1955) was brought up in Munich where his father owned an electro-technical plant. He did not do well at school, where he is said to have developed a suspicion of authority, but he did show an early aptitude for science and music. After several attempts, he finally entered the Swiss Federal Polytechnic School in Zurich where he graduated in 1900. In the same year he published his very first paper on the physics of fluids in drinking straws, in *Annalen der Physik*, in the same issue as Planck's paper on *quanta*. Einstein had a sound knowledge of surface tension effects; years later he assured the New York correspondent of the BBC [5], that water alone is sufficient for shaving facial bristles and that shaving cream is superfluous.

At first Einstein could not get an academic post, so in 1902 he took a job with the Swiss patent office. For an unwaged theoretical physicist this was an ideal day job. The patent office at that time would have had many patent applications for new inventions and discoveries because of the electrical revolution; he might have had to stretch his imagination to comprehend some of the more fanciful physics of backyard inventors. In later years he was always sympathetic and responsive to letters about new ideas in physics. He enjoyed the patent office job which gave him enough spare time to pursue his serious theoretical work.

In1905, without any of the resources of an academic affiliation, Einstein produced four epochal papers in *Annalen der Physik;* the **first** *'On an Heuristic View concerning the Production and Transformation of Light'* covered quantum theory and the photoelectric effect; the **second** *'On a New Determination of Molecular Dimensions.'*; the **third** *'On the Movement of Small Particles Suspended in Stationary Liquids Required by the Molecular Kinetic Theory of Heat'* concerned atomic size and dynamics; and the **fourth** *'On the Electrodynamics of Moving Bodies'* outlined the **Special Theory of Relativity.**

The work of a patent attorney involves assessing the true originality of an invention. This is very much concerned with the *'heuristic process'*, which is more simply known as the process of discovery. The best patent clerks and patent attorneys keep an open mind in their initial evaluation of a patent submission, with a mind-set that says *"This invention may sound crazy but let's suppose it's true."*

Einstein reviewed Lenard's paper on the *photoelectric effect* and Planck's on *black body radiation* in a similar manner, accepting their evidence at its face value and more confidently perhaps than they had themselves. From this evidence he was able to produce a truly original concept about the interaction of light with electrons, later recognised as the first scientific paper dealing with *quantum mechanics.*

Applying Planck's notion of *quanta* to electromagnetic radiation itself, instead of to the 'Hertzian atomic oscillators', as Planck had preferred, Einstein proposed that light is not a continuous wave but instead comes in tiny packets of energy or *quanta* [6]. Light of the same frequency that is of the same colour, comes in packets that have the same energy. So when a *quantum* displaces an electron from a metal atom, it gives the electron the same amount of energy and therefore the same velocity; as Lenard had observed.

More intense light of the same colour gives more *quanta* (called *photons* after 1926), which eject more electrons but at the same velocity. Changing the colour of the incident light changes its frequency and the amount of *quantum energy* carried by each *photon*. If light energy increases, then the *quantum (photon) energy* increases and therefore the velocity of the displaced electrons.

This theory accounted for Planck's and Lenard's findings very well. However it took some time for Einstein's interpretation of the *photoelectric effect* to become generally accepted and only after he had declared that a better understanding of light resulted from a fusion of wave and particle theories. Einstein received the Nobel Prize for this outstanding work in 1921, sixteen years after its first publication. In 1927 Heisenberg, Bohr and Born showed that particle and wave aspects of radiation could be resolved by the concept of *quantum statistics*; that the amplitude of a wave at each point represents the probability of a particle *(photon)* being close by.

In his paper on the *Special Theory of Relativity* it seems likely that Einstein may also have applied his patent attorney method to the *Michelson – Morley Experiment*. This experiment, using very accurate measurements, had shown that the *speed of light in space* was invariably constant and independent of the speed of its source or the speed of the observer; for example whether Michelson's interferometer was moving either towards the sun or away from it. Even Michelson had been at a loss to explain this phenomenal behaviour of sunlight.

In the *Special Theory*, Einstein accepted that in empty space, light travels at the same speed and that its velocity is unaffected by the speed either of its source or that of the observer. Moreover no signal or energy can travel faster than light. This acceptance posed a challenge to classical physics because it meant that if the *speed of light* was constant, space and time could not be. Newton's notion of *'absolute space and absolute time'* was no longer acceptable.

The replacement of *'absolute space and absolute time'* with *'relative space-time'* has peculiar consequences. To an observer, moving clocks run more slowly the closer they get to the speed of light. At the speed of light itself, they stop. Accelerated bodies also increase dramatically in mass as they approached the speed of light relative to the observer. These results appear quite fantastic and alien to everyday experience but have been confirmed by the behaviour of high speed particles accelerated to very high velocity, such as those in the 26 kilometre circular tunnel of the European Laboratory for Particle Physics near Geneva.

Einstein continued work on the implications of the *Special Theory* and came to realize the equivalence of mass and energy. From this he formulated the famous $E = mc^2$ equation in which E is energy, m is mass and c^2 is the square of the speed of light.

Although the *Special Theory* rendered *'absolute space and absolute time'* redundant, it still observed Newton's *Laws of Motion* [7].

Newton's definitions of *momentum*, *force* and *acceleration* are contained within $E = mc^2$, however the dimensions of space and time by which they are defined are no longer absolute. The amazing implication of this equation is that there is an enormous amount of energy locked up in every material thing [8].

The *Quantum Theory of Light* and the *Special Theory* also affected Maxwell's theory and its concept of propagation of electromagnetic waves through an all-pervasive *luminescent ether.* If light is composed of timeless particles of energy (*quanta*) and not waves, *ether* is as superfluous as shaving cream.

Einstein went on to produce his *General Theory of Relativity,* which was confirmed during a total solar eclipse in May 1919 and in 1921 he was elected to the Royal Society. In 1924 Einstein received a letter from Satyendra Nat Bose, a young Indian physicist at Dacca University about *Planck's Law* and the hypothesis of light *quanta,* which resolved some of the imperfections in Einstein's quantum hypothesis. Einstein generalized

Bose's method to develop a new system of statistical quantum mechanics: *Bose-Einstein statistics.*

In 1933 Einstein was deprived of his university position as professor of physics in the University of Berlin, a post which he had held since 1914 and of his German nationality by the Nazi regime. He moved permanently to Princeton where he was already a visiting professor of mathematics and theoretical physics. Up until this time Einstein had been a pacifist, but at the outbreak of the Second World War in Europe in 1939, he wrote to Franklin Roosevelt, the US President, to warn of Germany's capability of building an atomic bomb – a weapon that became possible as a consequence of his own theories.

The letter was vague, delayed in its delivery and is unlikely to have led to the creation of the Manhattan project which produced the first ever atomic bombs; however in future years Einstein felt himself to have been partly responsible. In 1944 Einstein contributed US$ 6 million, which he received from the auction of his original 1905 hand-written copy of the Special Relativity paper, to the allied war effort. A week before his death in April 1955, nearly ten years after the destruction of Hiroshima and Nagasaki, he signed Bertrand Russell's manifesto urging all nations to abandon nuclear weapons.

Although it took a while for his genius to be fully recognised, there is no doubt that for Albert Einstein, 1905, the year in which he published papers of unrivalled genius at the age of only 26, was equivalent to the years of 1665 and 1666 for Isaac Newton, when he was aged 24. Otherwise there are few similarities in personality or behaviour. Whereas Newton was uncommunicative and reclusive, Einstein was gregarious. According to the tactful CP Snow, as a young man he was full of '*animal spirits and vigour*'. Although amusing and a bit of a clown in company, he was also a trial for his wives and children.

Einstein was also very different from Newton in making his hypotheses and discoveries widely known by prompt submission and publication of papers; and when he became more famous, by arranging international conferences. As a patent clerk he would have been aware of the importance of intellectual property rights, more than any other physicist since Benjamin Franklin. He was always generous with his time and responded promptly to letters from young scientists with new ideas, even if they sounded *crazy.*

PHOTONS AND THE RETINA

Einstein's and Bose's statistical quantum mechanics and quantum electrodynamics seem far removed from biology and human physiology, however the response of the retina to light is the one of the more significant quantum electro-dynamic phenomena on the planet. With the unaided eye we are unable to see individual *quanta* (photons) and at a rough estimate only 10 percent of the photons that enter the eye at the cornea are detected by the rods and cones in the retina; the remainder are lost by absorption and scattering within the eye.

Fully dark-adapted, scotopic rods are extremely sensitive to very small amounts of light and about a million times more sensitive than cones in daylight. However even with full dark adaptation, several photons have to reach the retina for it to perceive a tiny flash of light. The emission of photons by a faint source is random and follows a statistical distribution; several may arrive in the eye close together and then for a brief period none. This quantum statistical nature of light accounts for loss of visual acuity and the difficulty of discerning fine detail in weak light, because insufficient sporadic photons reach the retina within the tenth-of-a-second time span that the eye needs to build up a coherent image. The amount of energy in a single photon of light is vanishingly small [9].

In a letter to an English physiologist about photons and the retina, Einstein whilst at the Institute for Advanced Study in Princeton in 1950 wrote:

"My knowledge about facts and theories concerning nerve processes is quite scanty but it seems to me.....(nerve) stimulation is an "all or none" effect. What happens there must be analogous to the functioning of a relay. The nerve process is initiated (but not supported) by the outside stimulus, the nerve process consuming energy which had been stored up in the nerve during the rest period."

"As far as stimulus through light is concerned it is created by the absorption of photons – probably by the absorption of one photon. This is an atomistic process, probably in the interior of the sensitive cell which initiates the nerve process as a kind of chain reaction."

Scanty though Einstein's physiological knowledge may have been, in the year 1950 this was a very perceptive résumé. The number of photons needed to activate a well-adapted rod receptor may be several rather than one; however, very small amounts of energy are involved and vision is one of the most sensitive of quantum electro-dynamic processes. It is now accepted that in a receptive rod, *photon energy* is captured by a molecule of *rhodopsin,* which before impact is in its *cis* form and curved. Capture of a *photon or photons,* raises an electron in the molecule to a higher quantum state, which straightens out the molecule to its *trans* form. This small amount of photon energy is insufficient to expel an electron from the molecule but is sufficient to change the electrical conductivity of the rod and to stimulate a nerve impulse in the retina.

Having done its work, *trans rhodopsin* decays rapidly and is then chemically recycled back to its *cis* form. As we all know, this process works better if we eat up our carrots. The different *photopsins* in colour-sensitive cones work in a similar manner to *rhodopsin* in rods.

RESUMÉ

The first plausible theories concerning vision and the nature of light did not arise until the 17th century when Kepler recognised the sensitive role of the retina. Descartes studied the anatomy of the eye, became aware from Snell of the mathematical behaviour of light and theorized about the psychology of vision. It was Descartes who first tried to establish the scientific link between light and vision. Newton's fascination with light resulted directly from his reading of Descartes' *Dioptrics* and Hooke's *Micrographia* and the doubts engendered in his mind by their theories.

It is claimed that Newton had a preference for a *corpuscular theory* of light, but his statements in *Opticks* were not dogmatic. There is no doubt about his interest in vision. In *Quaestiones* in the *Third Book of Opticks* he wrote *"Is not Vision performed chiefly by Vibrations of this Medium (ether) excited in the bottom of the Eye by the Rays of Light, and propagated through the solid, pellucid and uniform Capillamenta of the optick Nerves into the place of Sensation?"*

Huygens definitely regarded light as a wave phenomenon and so initiated the *'particle or wave'* controversy which undulated through the hypotheses of Young, Fresnel, Maxwell, Planck and Einstein.

In the first quarter of the 20th century, theoretical physics moved into a higher realm, especially with the development of *quantum theory.*

The *relativity theories* and *quantum physics* appeared to describe different universes from that of human biology. On the one hand astronomers measure distances to stars in hundreds of light years (one light year is almost 10^{16} metres or 10 million, million kilometres). On the other hand the size of an atom is of the order of 10^{-10} metres or a hundred millionth of a centimetre. And yet the *rhodopsin* molecules in our retinas are a link between these two universes. Widely different in scale, both universes have in common the constant speed of light **c** and the minute amount of energy in a single photon.

When we look at a faint star in the night sky our eyes can detect a very small number of photons, which have travelled enormous distances over long periods of time. The eyes do so by means of an extremely low-energy, photon-electron effect. Retinal rods link cosmology and particle physics. But *rhodopsin* molecules are not the only molecules in our bodies subject to the laws of quantum electrodynamics, all our molecules are. As Helmholtz had tried to establish for the laws of classical physics, quantum physics applies to all energy and matter, whether living or inanimate.

NOTES

1. Snell's law states that -
"When light passes from one medium (such as air) into another medium (such as glass), the *sine* of the angle of incidence bears a constant ratio to the *sine* of the angle of refraction." The constant is the refractive index of the medium and for air equals 1 and for glass equals 1.5.

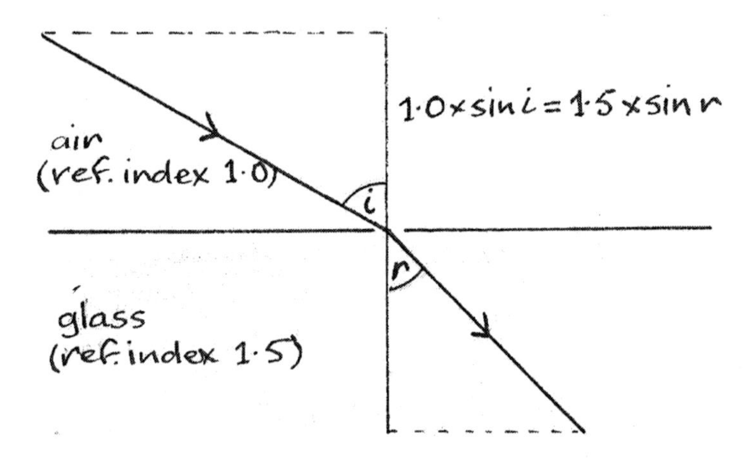

30. Snell's Sine Law of Refraction. The *sine* of the angle of the incident beam bears a constant ratio to the *sine* of the angle of the refracted beam.

2. A consequence of the wave theory and the discovery that light of different colours had **different wavelengths** travelled at the same speed in space, although at different speeds in optical media, was that they must have **different frequencies** of undulation (oscillation) that were inversely proportional to their wavelengths.

Frequency equals speed divided by wavelength

or $\nu = c / \lambda$ where ν is frequency, c is the speed of light and λ is the wavelength.

3. Refraction of light waves can best be explained by analogy with a formation of marching soldiers representing a beam of monochromatic, say red, light. When a wavelet of light passes from air into a denser medium its speed is reduced but red light remains red. In other words it maintains its frequency but reduces its wavelength, like a soldier shortening his stride. If, as the formation crosses a line at right angles to its march (as in A), the soldiers in each row shorten their stride, the column will slow but continue to move in a straight line. Likewise a beam of red light passing at right angles, from air into a denser medium such as glass will be slowed, reducing its wavelength, but will continue in a straight line.

If the formation needs to alter its direction, say to deflect its line of march at an angle to the right (as in B), the right marker, that is the soldier in the front row on the right, shortens his stride followed by the soldier to his left and so on all the way to the left marker, which will take an interval of time. The next row (wave) of soldiers then does the same and so on until the direction of the formation is angled to the right. The whole formation then marches more slowly but in a straight line in the new direction. In the same way each wavelet in an incident light beam at an angle, will be refracted according to Snell's Law, as its wavelength and speed is reduced on entering glass.

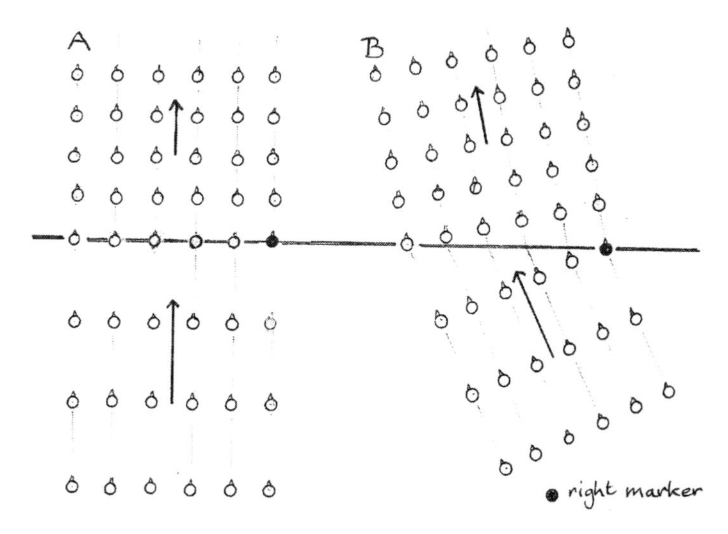

31. Refracting Formations. In **A**, soldiers in each row cross the line together at a normal angle; though they slow, they continue in a straight line. The rows of soldiers in **B** cross the line at an oblique angle so that those on the right shorten step before those on the left which angles the line of march.

Snell's Law of Refraction can be restated in terms of the speed of light - ***"The index of refraction is equal to the ratio of velocities of light in the first medium to that in the second."*** The velocity of light in air is 30,000 km/second and in glass is 20,000 km/second (slightly more for red and less for blue) giving glass a *refractive index* of 1.5.

4. Planck's ***Law of Black Body Radiation*** contained the supposition that electromagnetic radiation can only be emitted as ***quanta of energy,*** that is as multiples of an elementary energy unit E that equals hv, where h is ***Planck's constant*** and v is the frequency of the radiation in Hertz. Planck accepted the $E = hv$ equation with extreme reluctance because he had been compelled to use Boltzmann's statistical interpretation of the ***Second Law of Thermodynamics.*** Planck called it *"an act of despair....was ready to sacrifice any of my previous convictions about physics."* However his belief in the power of logical thought, based on experimental facts was so strong that he had proposed a law contrary to his previous convictions, because no other resort was possible.

5. Alistair Cooke was the BBC correspondent in New York for many years and interviewed Einstein for his regular British radio programme, *Letter from America.* After receiving shaving advice from the great man, Cook only used cheap disposable razors and never applied shaving cream again.

Both men suffered the posthumous misfortune of having their body parts stolen; Einstein's brain, for posterity by a Princeton pathologist and Cook's bones, to sell for bone grafts, by a New York mortician.

6. Applying Planck's equation $E = hv$ to electromagnetic radiation itself, instead of to the Hertzian atomic oscillators, as Planck had preferred. Einstein proposed that light is not a continuous wave but instead comes in packets of energy or *quanta.* Light of the same frequency v; that is of the same colour, comes in packets that have the same energy E. So when a *quantum* displaces an electron from a metal atom, it gives the electron the same amount of energy and therefore the same velocity; as Lenard had observed.

More intense light of the same colour gives more *quanta* (called photons after 1926), which eject more electrons but at the same velocity. Changing the colour of the incident light changes its frequency and the amount of energy carried by each photon. If light energy increases, then the *quantum (photon) energy* increases and therefore the velocity of the displaced electrons.

7. Newton's *Laws of Motion* are retained within $E = mc^2$ though the dimensions of space and time are no longer absolute and rigid, but are replaced by Einstein's elastic space-time, which may be 'tilted' or 'curved'.

Force = mass x acceleration - from Newton's Second Law
so **Force** (on a body starting at rest) = $\dfrac{\textbf{mass x speed}}{\textbf{time}}$

so **Energy = force x distance** = $\dfrac{\textbf{mass x speed x distance}}{\textbf{time}}$

Energy(joules) = $\dfrac{\textbf{mass (kg) x speed (m/sec) x distance (metres)}}{\textbf{time (sec)}}$

or **mass(kg) x speed (metres/sec) x speed (metres/sec)**

Where the maximum possible **speed (metres/sec)** is the ***speed of light c*** then **Energy (joules) = mass (kg) x c^2 (metres/sec x metres/sec)** and so

$$E = mc^2$$

In other words, the amount of condensed energy E within a mass m is the same as that required to accelerate it as quickly as possible, to go as fast as possible, to the top limit of the speed of light ***c***.

8. c^2 = **$3\text{x}10^8$ metres/sec x $3\text{x}10^8$ metres/sec**
or **$9 \text{ x } 10^{16}$ metres^2/sec^2**

so from $E = mc^2$ **; a mass of 1 kg contains $3 \text{ x } 10^8$ times $3 \text{ x } 10^8$** or **$9 \text{ x } 10^{16}$ joules of energy** and so
1 gram contains 90,000 million kilojoules.

The amount of energy in one gram of matter is sufficient to boil off all the water in about 30 Olympic size swimming pools.

9. *Planck's constant* (6x 10^{-34} joule.seconds) is extremely small; from the $E = hv$ equation, the energy of a single *photon* of visible light is also very small and is approximately 10^{-18} joules.

NEW RAYS, NEW ATOMS AND LIVING MOLECULES

After the experiments of Boyle and Hooke, physicists had been attracted to using vacuum chambers for their work so as to overcome problems with air and atmospheric pressure. The first **electric vacuum tubes** were made in Germany in 1858 and shortly after the English chemist and physicist Sir William Crookes designed a glass tube with an almost perfect vacuum. When two metal electrodes were sealed into the ends of a tube and connected to a source of high voltage, such as an induction coil, changes occurred as air was evacuated. At first the tube fluoresced but as pressure fell below 0.01 mm Hg it darkened, however green rays extended outwards from the negative electrode or cathode. Several laboratories in Europe and America researched the nature of these **cathode rays** and as mentioned earlier, JJ Thomson at the Cavendish Laboratory in Cambridge demonstrated in 1897, that they were caused by streams of electrons.

DISCOVERY OF X-RAYS

Another scientist who was determined to understand the nature of cathode rays was Wilhelm Conrad Röntgen (1845 – 1923), the professor of physics and head of the experimental physics department at the University of Würtzburg in Bavaria. His approach to physics was methodical and dedicated to making accurate observations and drawing valid conclusions from them. He was opposed to extravagant theorizing that exceeded the evidence.

Although Röntgen was born in Germany, his family moved to the Netherlands when he was young and so he received his early education in Utrecht; then graduated with a PhD in experimental physics from the University of Zurich. He was a very capable photographer and combined his love of hill walking with landscape photography, and also used photographic equipment in his laboratory.

Like Dalton, Röntgen was colour-blind. Although this is a disability of photopic vision it has some benefits for scotopic vision. Dark adaptation in colour-blind individuals occurs much more quickly than for those with normal vision

At that time photographic emulsion, a light-sensitive suspension of silver halides in gelatine, was coated as a film onto glass plates, to make photographic plates. These were held in light-tight cassettes until exposed to light within a box camera. In 1895 the emulsions were much less sensitive and therefore slower than now, so that camera exposure times were long. Placing fluorescent screens next to the photographic film during exposure increased the amount of light acting on the film and helped to reduce exposure times significantly. Röntgen used a fluorescent screen coated in barium platinocyanide in his laboratory for making photographs of his experiments. His research associate, Philipp Lenard, who was later to discover the *photoelectric effect,* modified an electric vacuum tube by inserting an aluminium window in its wall. Lenard detected some '*rays*' with a fluorescent screen, which he initially thought were cathode rays, outside the thin aluminium window.

On the evening of 8th November 1895, Röntgen decided to investigate Lenard's finding using a standard Crooke's tube, which had much thicker glass than Lenard's tube and no aluminium window. He covered the tube with black cardboard to block out light coming from cathode rays within it. As he was checking the effectiveness of the tube cover in the blacked-out laboratory he discovered that a fluorescent photographic screen, which had been left to one side, glowed very faintly when voltage was applied to the tube and ceased when it was switched off. He also found that this phenomenon was apparent at some distance from the covered tube. Röntgen then discovered that the effect was evident even when a thick book and then a wooden object were put between the activated tube and the screen.

Placing his hand between the tube and the screen he found that the soft flesh was transparent but he could identify shadows of the bones in his fingers. His conjecture was that the screen fluoresced because invisible rays from within the tube, different from cathode rays, were able to penetrate some solid materials. Röntgen's ability to see faint illumination in the dark better than most, must have been helpful in making the discovery.

Over the next eight weeks he investigated these invisible rays methodically, especially the way that metal objects absorbed or *attenuated* (thinned) the new rays. Platinum and lead attenuated the rays much more than zinc or aluminium. Röntgen showed that the absorption of rays by matter (*attenuation of the beam*) was dependent on the thickness, density and nature of the material as well as the voltage applied to the tube. Because he did not understand the nature of the new rays he called them ***x-rays***, ***x*** for unknown. However they soon became known as ***Roentgen rays***, a term he modestly avoided.

Röntgen quickly realized that photographic film contained in a light tight cassette could be used to make shadow images of metallic objects. He then went on to make an x-ray photograph of his wife's left hand on 22nd December. The rays were very weak and had to be focused on Frau Röntgen's hand for fifteen minutes to produce a shadow image.

Röntgen's report on *'A New Kind of Rays'* was submitted to the Physico-Medical Society of Würtzburg on the 28th December 1895. At the beginning of 1896 Röntgen sent copies of the *Proceedings* of the society to several scientists in Europe, including Lord Kelvin, JJ Thomson in Cambridge and Henri Poincaré in Paris. The newspapers soon learned of the discovery and by the end of January news had spread around the world, especially to the United States. Because vacuum tubes and induction coils were widely available it was not long before many others were able to perform x-ray experiments and to take x-ray photographs of patients with broken bones and gun shot injuries.

Thomas Edison built an x-ray machine in New Jersey within a few weeks of Röntgen's discovery and went on to manufacture machines with more powerful tubes, and better fluorescent screens that produced *'shadow photographs'*; also called *skiagrams, roentgenograms* or *radiographs,* with much greater detail. Edison then invented the *'fluoroscope'*, which allowed the operator to look at bones in hands and feet on a fluorescent screen without the need for photography. These machines fuelled a craze for x-rays, which were regarded by many ordinary people as magical.

Sir William Crookes, the inventor of the electric vacuum tube used by Röntgen, was at this time the president of the Society for Psychical Research and it is not surprising that some charlatans used the machines for spiritualist séances, suggesting that x-rays had supernatural powers. However the major use initially, was for the diagnosis of fractures and locating embedded metal foreign bodies, such as bullets and needles.

X-ray machines were used in British army hospitals in London in 1897 and then in surgical stations in South Africa during the Boer War. For the first time surgeons were able to *'see'* fractured limb bones and to set them so that they healed in satisfactory alignment. It was also possible to view fractured bones when they were immobilized by plaster castes, to assess healing. Unfortunately the early x-ray tubes were not powerful enough to make satisfactory examinations of the spine, hip and pelvis. Radiographs of these regions were faint even with exposure times of up to 30 minutes and burning of the skin associated with such long exposures was initially considered to be due to *'electrical effects'*. In spite of these limitations, radiography began to revolutionize the practice of bone and joint surgery and reduce the number of patients crippled by un-united or mal-united fractures.

Another revelation was the ability to view patients' lungs using an Edison fluoroscope. Compared with the pelvis, x-ray photographic exposure times for the chest were short. As early as 1897 physicians started to look at the different *'shadow patterns'* produced in their patients by lung diseases such as pneumonia, tuberculosis and emphysema. For the first time they were able to assess the extent of pulmonary diseases and to monitor their progress more accurately than with a stethoscope alone.

Detection of early tuberculosis, before clinical signs developed, combined with isolation of suspects in sanatoria, ultimately led to containment of the tuberculosis pandemic in Europe and North America even before satisfactory anti-TB drugs were developed.

In the early 20th century radiography transformed pulmonary medicine as it did bone and joint surgery.

Within a year of Röntgen's discovery of x-rays they were used to irradiate skin diseases with some initially beneficial results, however the doses given were uncertain and some patients were burned by overlong exposures. Ringworm, leprosy, acne, psoriasis and birth-marks were considered suitable for treatment but the most successful results were achieved for rodent ulcers (skin cell cancers). Some radiologists soon recognized the dangers of excessive x-ray exposures, both diagnostic and therapeutic, and x-ray injuries in 23 patients had been reported in the medical literature by January 1897. In 1898 the English Röntgen Society established a committee to enquire into '*the alleged injurious effects of Roentgen rays*', in response to a warning from Röntgen himself, who was extremely careful and insisted on adequate lead protection for all those working with x-rays.

Yet in spite of warnings, many physicians and radiographers used x-rays without protection and regarded the dermatitis and ulceration of their hands that resulted as acceptable. Some of the ulcers were malignant and required amputations of fingers and then hands. One of the most unfortunate was one of Edison's engineers who tested all new fluoroscopes made in the factory by viewing his own hands. He noticed burns to his face and hands and then developed rapidly spreading, malignant ulcers that required amputation of both arms; he died in agony in 1904. The paradox was that x-rays could be used to treat cancer but could also cause it.

It may seem that Röntgen's discovery of x-rays was serendipitous. Other researchers had the same vacuum tubes, induction coils and photographic techniques available to them and could have made the discovery; however his determination to investigate cathode rays and photographic expertise were crucial. Blackening of photographic film '*exposed*' in light-tight cassettes had been noticed in Crooke's own laboratory but had been blamed on faulty photographic film.

Röntgen had a prepared mind and the specialist skills necessary to discover ***x-rays***, to invent ***radiography*** and to found the science of ***radiology***. The first Nobel Prize in Physics in 1901 was well deserved but for many years after the discovery of x-rays their true nature remained unknown.

At that time x-rays could be detected in one of three ways –

1. ***Fluorescence of crystals*** (barium platinocyanide or calcium tungstate).
2. ***Exposure of photographic film*** in light-tight (radiographic) cassettes.
3. ***Electrification of air in an ionisation chamber.***

Air is normally a very poor conductor of electricity but x-rays ionise the oxygen and nitrogen molecules and so increase electrical conduction. Electric charge produced in an ionisation chamber is proportional to the intensity of the x-ray beam and can provide a quantitative measure which until 1957 was the ***roentgen*** [1].

Early attempts to reflect, refract or diffract x-rays, like other electromagnetic radiations, were unsuccessful. Then in 1912, it was discovered that x-rays could be diffracted, like other electromagnetic radiations, by crystals of copper sulphate and had very short wavelengths of around 10 picometres (10^{-11} metres) and very large *photon energies* of 100 kilo-electron volts. After this, the dual wave-particle *(photon)* hypothesis for electromagnetic radiation became increasingly important for understanding the physics of x-rays.

The discovery that x-rays could be diffracted by crystal structures and that the three-dimensional arrangements of atoms within them could be determined by analysis of x-ray diffraction photographs was the start of the science of ***x-ray crystallography,*** which was pioneered by William Henry Bragg (1862 – 1942) professor of physics at Leeds University and his son William Lawrence Bragg ; both received the Nobel Prize for Physics for their x-ray work in 1915. They published *X-rays and Crystal Structure* in the same year. The technique became a powerful tool for analysing the molecular structure of both organic and inorganic molecules.

RADIOACTIVITY

Antoine-Henri Bequerel (1852 – 1908) was the hereditary professor of physics in Paris, at the Museum of Natural History, in succession to both his grandfather, Antoine-César (1788 – 1878), who had worked with Ampere on the relationship of electricity to heat, light and chemical forces; and his father Alexandre- Edmond Bequerel (1820 – 1891), who had studied electromagnetism and optics, was also famous for his work on the phosphorescence of uranium compounds, luminescence and the photoelectric cell.

Henri Bequerel studied physics and then engineering. His early research in physics was optical, concerned with phosphorescence and the rotation of plane-polarized light by magnetic fields and by crystals. On the death of his father in 1891 he became professor of physics in Paris and two years later chief engineer to the National Administration of Bridges and Highways; and thought his research days were over.

Bequerel learned of Röntgen's discovery at the Academy of Sciences on 20th January 1896 and discussed it with Henri Poinacré. He wondered whether luminescent crystals might emit x-rays as well as visible light. He knew about different luminescent crystals from his father's research. On 24th February he announced to the Academy that sunlight shining on crystals of uranium sulphate produced visible luminescence, and like x-rays, had also exposed a photographic plate in a light-proof envelope.

At first Bequerel conjectured that ultraviolet rays in sunlight caused the crystals to emit invisible penetrating radiation similar to x-rays, as well as visible light. One week later he reported his puzzlement at finding that the photographic plate was blackened even when the uranium sulphate crystals were kept in the dark and were not luminescent. Moreover the penetrating rays persisted unchanged when the crystals were either heated or dissolved in water. Bequerel then studied other uranium compounds, whether luminescent or not, and found that they all produced rays; the most penetrating from a disc of pure uranium.

By May he was able to prove that the power of emitting penetrating rays was a property of uranium and not directly related to luminescence. He also found that like x-rays, they caused electrical changes in the surrounding air, which were due to ionisation. The emitted radiations became known as ***uranium rays*** or ***Bequerel rays***.

In the general excitement over x-rays, Bequerel rays were largely ignored, except by a Polish postgraduate looking for a rewarding research project. Marie Curie (1867 – 1934) was born Marie Sklodowska in Warsaw and became a governess before going to the Sorbonne to study mathematics and physics in 1891. After graduating, she married Pierre Curie (1859 – 1906), a physicist who taught at the Municipal School of Industrial Physics and Chemistry in Paris, where she was allowed to use a small back room as a laboratory for her postgraduate study of *Bequerel rays*.

Several years earlier, Pierre and his brother had invented a sensitive electrometer for measuring very small electric currents. Marie used the electrometer to measure the flow of charge in air that had been electrified (*ionised*) by uranium rays. She confirmed Bequerel's findings by showing that the electrical effects were constant whatever the state of the uranium and that they were directly proportional to the amount of uranium present. She concluded that the emission of Becquerel rays was an intrinsic property of uranium atoms. Marie and Pierre then tested other elements to see if they ionised air in the same way as uranium. In 1898 they discovered that thorium also emitted *Bequerel rays* and coined the word ***radioactivity*** to describe the phenomenon. Uranium and thorium were the first to be called ***radioactive*** elements. Bequerel and the Curies may have had an intuition that the electrical effects induced in surrounding air by uranium and thorium were linked in some way to Hertzian electromagnetic radiation, later called wireless ***radio** telegraphy*.

In her research Marie had observed that pitchblende, a complex uranium ore, contained elements that were more radioactive than uranium itself. The quantities were very small and it was necessary to process large amounts of pitchblende to obtain sufficient material for experiments.

Ten tonnes of the ore, containing uranium oxides and impurities, were stored and then processed by Pierre and Marie in a disused medical dissecting room behind the Municipal School, which she called the *'miserable old shed'*.

Using a sequence of chemical separations combined with radioactivity measurements, Marie and Pierre first discovered a highly radioactive element chemically similar to bismuth, which they named ***polonium*** in honour of Poland. At that time Marie's home country had been annexed and partitioned between Russia, Prussia and Austria-Hungary and she used the name of the new element to publicize Poland's plight. It is ironic that polonium-210 was used more than a century later, by persons unknown, to poison Russian secret agents [2]. Polonium is so very radioactive with a half-life of only 138 days that Marie was unable to isolate it in a pure form, despite determined attempts.

Six months later, in December 1898, the Curies announced their discovery of another radioactive element with chemical properties similar to barium, which they named ***radium,*** from *radius* the Latin word for ray. Radium-226 has a much longer half-life than polonium and after three years of intensive laboratory work Marie was able to isolate 6.5 milligrams of pure radium chloride from two tonnes of pitchblende.

Working in collaboration with the Curies, Henri Bequerel continued his research of radioactive radiation using radium instead of uranium. Initially he thought the radiations were similar to x-rays, but found they did not penetrate material in quite the same way. Following JJ Thomson's discovery of electrons, experiments with radium in Paris and Cambridge showed that radium rays were partially composed of high-speed electrons. These accounted for some of the ionisation effects detected with Pierre Curie's electrometer. In 1903 the Curies shared the Nobel Prize for Physics with Henri Bequerel for their work on radioactivity. Pierre was then appointed as a professor to the Faculty of Science at the Sorbonne.

Radium salts were produced in larger quantities at a factory in the eastern suburbs of Paris after1904 and its magical qualities were widely proclaimed, much as x-rays had been eight years earlier.

The major purchasers of radium from the Paris factory were physics laboratories in Vienna, Berlin, Cambridge and Montreal who wanted to perform experiments with radioactivity. Pure radium salts were extremely expensive because even the most efficient extraction methods of richer ores from the Belgian Congo, yielded only one gram of radium bromide per tonne of uranium ore.

Radium did not offer an alternative to x-rays for making diagnostic radiographs; however it was used in luminous paints for wristwatches, jewellery and instrument dials. For a short time it became an additive in cosmetics, toothpaste and cocktails, because of its supposed curative properties. These uses were soon prohibited after its dangerous effects on health were discovered. More than a hundred dial painters, who used their lips to shape their brushes, died of radium poisoning in later years.

In 1901 Henri Bequerel experienced a localized skin burn on his chest, after carrying a vial containing radium in his coat pocket, which was clinically identical to x-ray dermatitis. The Curies then loaned a radium cell to a dermatologist who used it to irradiate skin cancers and he achieved some successful results. Bequerel reported these medical activities in his Nobel Prize Lecture and as a consequence, a radium treatment laboratory and clinic were established in Paris in 1906.

It soon became clear that some tissues were more affected by radiation than others. Cells that multiplied rapidly such as those in skin, mucous membranes lining the airways and digestive tract, and active bone marrow, were damaged or killed by radiation much more easily than those in muscle or fatty tissues. With exceptions, rapidly growing malignant tumours were generally more sensitive than slow growing benign tumours.

After Pierre's death in a traffic accident, Marie continued their research and then succeeded Pierre as professor at the Sorbonne. Marie Curie was determined to establish a laboratory for the study of radioactivity and its applications in physics, chemistry, biology and medicine. She was awarded the Nobel Prize for Chemistry in 1911.

The University of Paris and the Pasteur Institute joined forces to establish the Radium Institute, which was built close to the Sorbonne in 1914. It had two sections; the Curie Laboratory headed by Marie for research in physics and chemistry and the Pasteur Laboratory directed by a cancer physician, Claudius Regaud (1870 – 1940), for radiobiological and medical research. The two of them worked together until her death in 1934. His aim was to develop the *'scientific treatment of cancer'* by integrating basic radiobiological research with applied research in the clinic. The clinical science of ***radiotherapy,*** also known as ***radiation oncology,*** was established.

Although x-rays were better for diagnostic radiography, radium applicators gave much more precise radiation doses and were more suitable for radiotherapy in the early days. Successful radiotherapy depends on giving the maximum effective dose to the entire cancer whilst at the same time minimizing the dose to healthy tissues surrounding it. Research showed that dividing treatments over several sessions (*fractionation*) could reduce damage to normal tissues and still treat the tumour.

During the First World War radium research was suspended. Marie Curie became Director of the Red Cross Radiological Service and decided to equip motorcars with mobile generators and x-ray machines. These were used on the battlefields of northern France to help French and Belgian military surgeons remove bullets and shrapnel. She was able to equip twenty cars, which became known as '*petit curies*'. With her daughter Irène she escorted the cars to the front, making sure the x-ray machines worked properly; she often assisted surgeons to identify fractures, bullets and shrapnel fragments. During the war she used the Radium Institute as a radiography school and trained over 150 radiographers in x-ray techniques.

After the war the Radium Institute returned to its planned research programmes and Claude Regaud established a clinic close by for studying the radium treatment of patients.

He worked closely with Marie and they were able to design applicators and needles to position radium effectively so as to treat cancers in different parts of the body: the skin, breasts, orbits, tongue, jaws, thyroid, uterus, cervix and lymph glands.

Treatment doses of radiation could be closely controlled by application times and by measurement of the strengths of the sources, for which Marie took personal responsibility, well aware of the harm caused by mistakes. The unit of radioactivity, the ***curie*** was named for Marie and Pierre, however it is no longer used officially and has been replaced by the ***bequerel***[3].

Marie Curie measured and certified the amounts of radium and its radioactive daughters, in treatment applicators used on patients in the Paris clinics as well as consignments to hospitals and laboratories elsewhere in Europe, which must have taken a significant amount of effort. Many years later the certificates revealed traces of radium where she had held and signed the documents, indicating that she had been heavily contaminated during her work. Radium absorbed into the body, like other heavy metals such as lead and strontium, becomes concentrated in bone, which surrounds the red marrow where blood cells are formed. The resulting irradiation of the marrow can either induce leukaemia or kill blood-forming tissues so as to cause aplastic anaemia, and it was from this that Marie Curie died in 1934.

In 1991, the Claudius-Regaud Hospital, a modern cancer clinic, opened next door to the Curie Institute whose vocation continues to be research, teaching and cancer treatment. Marie Curie research centres, of which there are many, remain dedicated to the exchange of knowledge and research, between the laboratory and the clinic.

Early attempts by Henri Bequerel and the Curies to understand how or why certain elements were radioactive were unsuccessful. However other laboratories in Europe and North America were also studying radioactivity with radium supplied by the Curies. The break through came in Montreal where the newly appointed professor of physics at McGill University, Ernest Rutherford (1871 – 1937), detected radiation that had spread into corners of the radium lab. Rutherford realized that the *'emanation'* was caused by a radioactive gas released by radium. He also found that the radioactivity of thorium diminished with time.

Ernest Rutherford was born in a coastal town in the South Island of New Zealand where his family had a small farm. His father was a handy man who started a utility business repairing electro-mechanical gadgets, which were recent and novel imports at that time. Ernest became interested in science whilst at school and won scholarships to study at Canterbury College, Christchurch, where he graduated in mathematics and physics, and then in 1895 to a research fellowship at the Cavendish Laboratory in Cambridge. At Christchurch he had developed a detector of very high frequency *Hertzian (radio) waves,* and continued this work in Cambridge where in February 1896 he demonstrated radio transmission over a record distance of several hundred metres, even through brick walls.

Professor JJ Thomson realized his exceptional ability and persuaded Rutherford to work on the electrical conduction of gases, leaving the glorious achievement of long distance wireless telegraphy to Guglielmo Marconi. Rutherford designed an ingenious ionisation chamber to study electrical conduction through gases when high voltages are applied; normally gases are very poor conductors. Very soon after Röntgen's discovery, he demonstrated that x-rays could induce electrical conduction in gases by ionisation and that *Bequerel rays* from radioactive elements also ionised the air surrounding them. By 1898 he had discovered two types of radiation from radium, which he called α and β rays. As there was no position for him in Cambridge, he accepted the professorship at McGill University where the laboratories were well equipped, and wrote to his fiancé: *"I am expected to do a lot of work and to form a research school in order to knock the shine out of the Yankees!"*

Following his discovery of α and β rays and the radioactive *emanation* in his new laboratory, he became more determined to understand the nature of radioactivity. Rutherford, with the help of a young English chemist, Frederick Soddy (1877 – 1956) who came to McGill in 1901, came to realize that radioactive elements changed into other radioactive elements and emitted radiant energy of three different sorts as they did so. They also formulated the concept of radioactive half-life to indicate the rates of transmutation of each element.

The half-life of a radionuclide (radioisotope) is the time taken for its radioactivity to decay to half of its original value.

Radioactive decay is an exponential process; for example after two half-lives, activity falls to a quarter and after ten it falls to nearly a thousandth of the original. Radium which has a long half-life loses only one percent of its radioactivity after 25 years.

In 1905 Rutherford reported the chain of transmutations of radium and its daughter elements; which were labelled Radium A to F through to stable lead, in the *Philosophical Transactions of the Royal Society* [4].

Two years later in Manchester, Rutherford proved that ***α rays*** were particles with four times the mass of a hydrogen atom and were positively charged ***helium nuclei*** ($^4He^{++}$); helium atoms that had been stripped of their electrons. Alpha particles were barely able to penetrate tissue paper but did cause scintillation in sodium iodide crystals. By 1899 both Henri Bequerel and Rutherford had confirmed that ***β rays*** were negatively charged particles, which had a mass less than 1/1000th of the mass of a hydrogen atom and were ***high-speed electrons*** **(e^-)**, with a depth of penetration in normal tissue of a few millimetres.

Rutherford found that ***γ rays*** had no charge or mass, were ***high-energy rays*** and passed through the body with partial attenuation, similar to penetrating x-rays; later shown to be high frequency ***electromagnetic waves***, or high-energy ***photons***.

In 1904 Rutherford published *Radioactivity*, his first book, and in 1908 was awarded the Nobel Prize for Chemistry "*for investigations into the disintegration of elements and the chemistry of radioactive substances.*" Rutherford remarked that the most rapid transmutation had been of him-self; changed from being a physicist into a chemist.

HUNTING THE ATOM.

Up until the late 19th century, physicists were not especially interested in matter on a minute scale. Most scientists assumed that matter was made up of infinitesimally small, indestructible atoms and were prepared to accept Dalton's theory as a convenient conjecture for chemistry and Berzelian notation as no more than a hypothetical language for describing chemical reactions. Even if atoms really existed most scientists thought that they would be too small to be detected.

For physicists, Mendeleyev was a chemist and the *Periodic Table of Elements*, though intriguing, was within the sphere of chemistry not physics; atoms belonged to chemists! The partition of the two sciences had been accepted by the British Association for the Advancement of Science since1834 and had been reinforced by the divergence of industrial chemistry and electromechanical engineering as the industrial revolution developed. Then at the turn of the century three physicists, the Curies and Rutherford, had found new chemical elements and filled some spaces in Mendeleyev's Table and two physicists were awarded Nobel Prizes for Chemistry!

After Dalton and Berzelius, chemists regarded the atoms of different elements as rather like tiny snooker balls of different colours and atomic weights, which were held together by electric forces to form molecules. When JJ Thomson discovered the electron he turned the atom from a *'snooker ball'* into something more like a *'currant bun'*, with negatively charged electrons stuck like currants in the positively charged dough of the bun. Lord Kelvin likened Thomson's atom to a *'plum pudding'*.

In 1907 Rutherford accepted the invitation to become the professor of physics at Manchester University so as to become closer to other research centres in Europe. Rutherford seemed to have an inspired intuition for solution of problems. According to CP Snow, the Cambridge physicist and novelist, he *"seemed scarcely ever to have tried a problem which wouldn't go"* and was the most outstanding experimental physicist since Faraday. Rutherford was also suspicious of theoretical physicists who didn't do their own experiments but used other peoples' results.

Rutherford took possession of Thomson's new atom and ran with it; there was to be no partition between physics and chemistry at the atomic level. He was determined to understand the inner structure of atoms and devised instruments and a sequence of ingenious experiments towards that goal. In Montreal Rutherford had observed that a beam of α *particles* became diffused on passing through a thin sheet of mica. He and Hans Geiger, his assistant in Manchester improved their ionisation detector, which was a fore-runner of the Geiger-Muller counter, to measure high-energy radiation. They also re-designed a scintillator crystal so that it could be used to detect α particles more effectively.

One of Rutherford's students using a scintillation detector discovered that a small proportion of α *particles* directed at a thin of leaf of gold bounced straight back; though most particles passed straight through or were deviated slightly. Rutherford was very surprised and said that it was as if one had fired a naval gun at a sheet of paper and the 15 pound shell had bounced right back. In 1910 he deduced from this that the mass of an atom is concentrated in a small, positively charged nucleus at the centre of the atom, with electrons orbiting around it. The Rutherford nuclear model was likened to the solar system with electrons orbiting the nucleus like planets around the sun.

It became clear to Rutherford from his study of radioactivity that there was an enormous amount of energy locked up in atoms and that α, β and γ rays were emitted by the nuclei of disintegrating atoms of radioactive elements. He also knew that each flash of light in a scintillation detector represented one α particle and one nuclear disintegration; so that it *was* possible to detect the decay of a single atom.

After long and tedious experiments counting scintillations in a blacked-out laboratory and measuring the proportionate loss of weight of radioactive elements, he was able to calculate the weight of individual atoms and their number in a gram-atom of material [5]. Rutherford also proposed a method of geological dating by measuring the radioactive decay of uranium-235 in rocks.

In almost every respect Niels Bohr (1885 – 1962) was the antithesis of Ernest Rutherford in background and personality. He came from a liberal and cultured Danish family devoted to the arts as well as science. Bohr's father Christian, the professor of physiology in Copenhagen, did much of the research to unravel the mysteries of respiratory physiology. His mother belonged to a wealthy Jewish family, prominent in banking, politics and philosophy. Diffident and always courteous in manner, Bohr spoke discursively in a whisper, whereas Rutherford's impatience and loud voice could shake laboratory instruments and ruin experiments.

Bohr studied mathematics and physics in Copenhagen, completed his doctoral thesis on the electron theory of metals in 1911 and then went on to Cambridge to continue his research on electrons with JJ Thomson. For some reason Thomson took no interest in Bohr's work; so after waiting for six months, Bohr travelled to Manchester and explained tentatively to Rutherford that he had been having a bad time at Cambridge. CP Snow maintained that behind the tobacco smoke, bluff colonial manner and caustic one-liners, Rutherford was really a sensitive and perceptive individual.

Whether true or not, Rutherford liked Bohr at once, listened patiently to him and formed a high opinion of Bohr's capabilities as a physicist. Between them they started to sort out the atom; Rutherford attacked the nuclei of different elements with α *particles* and Bohr contemplated the arrangement of electrons in atoms.

After the First World War, during which he helped to develop hydrophones to detect enemy submarines, Rutherford became Director of the Cavendish Laboratory. He found that bombarding nitrogen with α *particles* produced oxygen and positively charged hydrogen ions which must have been split off from the atomic nucleus. He called them ***protons,*** from the Greek word meaning first. It became clear that the hydrogen atom had a nucleus of just one proton 'circled' by one electron.

As far back as 1911 Frederick Soddy, the chemist who worked with Rutherford, had noticed that elements could come in different versions and have identical chemical properties but different atomic weights; for example the lead decay products of radium - lead-214 (radium B), lead-210 (radium D) and stable lead-207. He called them ***isotopes,*** which means being in the same place, that is in the *Periodic Table.*

It was not clear why many elements had different isotopes until 1932 when Rutherford's colleague at Cambridge, John Chadwick (1891 - 1974), discovered that atomic nuclei contained other particles of the same mass as protons but no charge, which he called ***neutrons.*** Nuclei with the same number of protons could have different numbers of neutrons and therefore represent different *isotopes* of the same element, for example lead [7].

Most combinations of protons and neutrons are stable but those of radioactive isotopes are unstable and therefore disintegrate. When they lose protons, unstable atoms decay into other elements. Nearly all the weight of an atom is in its nucleus and so the nuclei of isotopes are known as ***nuclides;*** the nuclei of radioactive atoms are known as ***radionuclides.***

A simple working hypothesis is that nuclei of atoms larger than hydrogen are composed of protons and neutrons; and that their combined numbers are indicative of the ***atomic weight*** of an isotope of an element. The ***atomic number*** [Z]of an element is the number of protons in its nucleus. Different isotopes of the same element have different numbers of neutrons and so different atomic weights but the same number [Z number] of protons and 'orbiting' electrons; therefore the same chemical properties and the same place in the *Periodic Table.* Although particle physicists now postulate that protons and neutrons are composed of elemental particles known as *quarks* bound by *strong* and *weak nuclear forces;* the *proton-neutron nucleus* concept is still useful in practice.

Niels Bohr continued his quest with electrons, knowing as Rutherford knew, that the planetary model of the atom; of negatively charged electrons orbiting the positive nucleus, could not obey Faraday's laws and endure. According to the laws of classical physics, electrons that continuously change velocity in circular orbits would emit electromagnetic radiation; lose energy, spiral into the nucleus and collapse the atom. As this did not happen, Bohr postulated that electrons must exist in stable *'energy states'* within the atom that did not require them to accelerate or to radiate. The only evidence for this arbitrary proposal was that atoms endured and didn't collapse.

Nineteenth century spectroscopists and chemists had discovered that chemical elements heated in a Bunsen flame emitted light of particular frequencies, *spectral lines,* which were characteristic for each element. But they didn't know why. Bohr used Einstein's quantum analysis of photon-electron phenomena and *Planck's Constant* ***(h)*** and was able to calculate discrete energy levels for a single electron in a hydrogen atom. By 1913 he came to realize that transition of an electron from one energy level to another level within the atom, caused by a *quantum (photon)* of light energy, accounted for the observed spectroscopic frequencies of light waves ***(v),*** or quantum energies of photons ***(hv),*** which were either absorbed or emitted by hydrogen atoms.

The energy levels are very precise so that a photon absorbed into an atom must have exactly the right *quantum* of energy (***hv***) to lift an electron to a higher energy level within the atom. And a photon with the same *quantum energy* (***hv***) or a light wave of exactly the same frequency (***v***) is emitted when an electron falls from the high energy level back to its original level

Photons with *quantum* energies that don't match electron energy levels will not be absorbed and can't be emitted [6].

Bohr returned to Copenhagen in 1916 and a few years later the Danish government established the Institute of Theoretical Physics of which he became the director. He called the possible electron energy states within an atom 'possible orbits', which was misleading because it retained Rutherford's notion of a planet-like system a with electrons circling the nucleus, whereas Bohr regarded the energy states in a stable atom as static, unless disturbed by quanta. Subsequently the electron energy states became to be regarded more like shells around the nucleus; like the layers of an onion. When Einstein learned of Bohr's findings, he was delighted and said

"*Then this is one of the greatest discoveries.*" He realized that Bohr's use of Planck's quantum for electrons matched his quantum theory of light particles (*photons*). It explained photon-electron interactions and Planck's original '*black body*' experiments, the subject of Einstein's 1905 paper. In 1916 Einstein introduced quantum probability by observing that some spectral lines for hydrogen were stronger than others and that some electron quantum jumps were more likely to happen than others. By 1920 the '*principle of complementarity*', which proposed that the wave and *quantum (photon)* theories of light were not mutually exclusive but *complementary*, began to gain acceptance. Einstein was at last awarded the Nobel Prize for Physics in 1921 for his work on light quanta published in 1905.

Whilst Rutherford and Chadwick continued to probe the atomic nucleus, other theoretical physicists wanted to know how the electrons were arranged in atoms and what governed their energies and motions using the new quantum theory.

One of the first to use Bohr's quantum concept to explain the atomic arrangement of electrons was the German physicist Walther Kossel (1888 – 1956) who became professor of physics at Kiel University. In

1916 Kossel observed that the first three noble gases, so called because they were chemically inert and would have nothing to do with ordinary elements; had atomic numbers and so electron numbers, which increased by 8; helium-2, neon-10, argon-18. Kossel surmised that the electrons in these inert elements were in *'closed shells'*, two electrons in the first shell [K] and eight in the second [L]. Hydrogen-1, fluorine-9 and chlorine-17 each have one electron less in their outermost shells than helium, neon and argon and are very active chemically. Fluorine and chlorine are strongly electronegative non-metals, which readily form salts with metals. Elements with one electron more; lithium-3, sodium-11 and potassium-19 readily form soluble hydroxides in water and are known as alkaline metals. Kossel's discovery began to explain the arrangement of the elements in the *Periodic Table*. It also accounted for the chemical combination of elements to form molecules, because atoms combined so as to fill the spaces in their outer electron shells.

ANATOMICAL ATOMS AND MOLECULES

This new model of the atom proved to have significant explanatory power, not only for atomic physics and chemistry, but also for biochemistry and biology. The atoms of just four elements - oxygen, carbon, hydrogen and nitrogen - together account for 96 percent of the weight of an adult human body; 70 percent of which is water. The metals – calcium, potassium, sodium, iron, magnesium and iodine make up 2.5 percent; phosphorus, sulphur, chlorine and traces of other elements account for the remaining 1.5 percent of body weight [8].

Although **oxygen** is the largest component by weight there are more than twice as many hydrogen atoms as oxygen atoms in the body; carbon atoms are the next most numerous. The gaps in their outermost electron shells determine the valencies of elements; the ratios in which their atoms combine with others to form molecules. A **hydrogen** atom has a single electron, whereas an oxygen atom needs two to complete its outer shell; by combining to form **H_2O,** these atoms complete their outer shells.

Sodium with only one free electron binds well to chlorine, which requires just one electron to complete its outer shell and so form common salt, **NaCl**. Although nineteenth century chemists such as Berzelius knew elements had particular valencies they did not understand why. The quantum atomic theory provided an explanation. The theory also showed why carbon is such a gregarious element.

Carbon occupies a central place in the Periodic Table of elements being neither electropositive like metals nor electronegative like chlorine and other halogens. It more often shares its four outer electrons with other atoms rather than surrendering them or capturing electrons from others. Carbon atoms can form molecules containing single, double or treble bonds by sharing electrons.

Ethane (**$H_3C\text{-}CH_3$**) is a single carbon bonded molecule.
Ethylene (**$H_2C{=}CH_2$**) has a double carbon bond.
Acetylene (**$H\text{-}C{\equiv}C\text{-}H$**) has a triple carbon bond.

The flexible bonding of carbon atoms to one another and to other atoms makes possible the formation of long chained organic molecules. Molecules of different **amino acids** which each contain an amine group **$-NH_2$** and an organic acid group **$-COOH$,** can link together to form **protein** chains containing thousands of amino acids. Approximately three-quarters of body solids are proteins, which perform an enormous variety of functions; **structural proteins** define the anatomy of the human body from bone marrow to outer skin; **enzymes** transport and transform other molecules; **hormones** regulate and **nucleoproteins** organize cell structure and govern cell division; other specific proteins perform respiratory, metabolic and immunological functions.

Anatomically the most significant proteins are elongated **fibrous proteins** such as **collagen**. This is the basic structural protein of bone, tendons, ligaments, cartilage and connective tissue and has enormous tensile strength. **Collagen** and **elastin** fibres in the walls of arteries, veins, tendons and skin give these structures the support and pliability they need to withstand powerful, changing locomotive and haemodynamic forces. **Compact bone** is composed of a tough collagen matrix strengthened by hydrated calcium phosphate crystals **$Ca_3(PO_4)_2 \cdot (H_2O)_x$**, which give bone its compressional strength. The collagen fibres extend along the lines of tensional force, like steel cables in reinforced concrete, to give bone its tensile strength. **Actin** and **myosin** are the contractile proteins of muscle.

Lipids are organic molecules characterized by chains of carbon and paired hydrogen atoms **$-CH_2$-CH_2-**. Stores of **neutral fat** (triglyceride) in adipose tissue are the most evident body lipids, however **phospholipids** help to form cellular membranes throughout the body and **myelin,** which surrounds and insulates neuronal axons, is present in the brain and peripheral nerves.

QUANTUM UNCERTAINTY

Although Bohr had explained the spectral lines for hydrogen mathematically he had not been able to extend quantum theory to explain the spectra of other elements. In the 1920s groups of mathematicians and physicists in Copenhagen, Göttingen, Berlin, Zurich, Vienna and Cambridge began to unravel the complexities of sub-atomic particle physics. Of the many Nobel Prize winning physicists involved, possibly the three most well known are Heisenberg, Schrödinger and Dirac.

Werner Heisenberg (1901 – 1976) was born in Munich where his father was professor of Greek. Towards the end of the First World War he worked as a farm boy in Bavaria and managed to avoid starvation. After Germany's defeat, at the age of 18, Heisenberg joined the Young Bavarian League, took part in the military suppression of Bavarian communists and spent his spare time playing chess. From 1920 he studied mathematics and physics in Munich, then Göttingen and later worked with Bohr in Copenhagen. With the help of mathematicians in Göttingen he applied matrix algebra, a kind of chessboard mathematics, to the electron problem. By 1926 he was able to explain the spectral lines of sodium, then of other elements. This mathematical approach, which became known as ***quantum mechanics*** provided a system to analyse the spectra of atoms and molecules.

Heisenberg realized in 1927 that determinations of the position and momentum of a particle as small as an electron are subject to variation and that this indeterminacy is mathematically implicit in the equations of quantum mechanics. This notion is often known in English as the ***Uncertainty Principle*** but should probably be better translated as the *Indeterminacy Principle* which states that:

"The more precisely the position of a particle is determined, the less precisely its momentum is known, and conversely."

This means that such a thing as an electron with a precise momentum in a precise position does not exist. Heisenberg had shown that the laws of classical physics did not work at the atomic level. He also believed that it was impossible to describe atomic events in plain language and said (translated from German) *"I think modern physics has definitely decided in favour of Plato. In fact the smallest units of matter are not physical objects in the ordinary sense; they are forms, ideas which can be expressed unambiguously only in mathematical language."*

In 1932 Heisenberg was awarded the Nobel Prize for Physics and during the Second World War headed the German nuclear weapons project. Heisenberg was unsuccessful in making an atom bomb during the war, because most of the best nuclear physicists in Germany had been Jewish and had been removed by Hitler. After the war Heisenberg became director of the Max Planck Institute. He supported the peaceful use of nuclear energy but opposed the use of nuclear weapons.

The seventh Duc de Broglie, Louis-Victor Pierre Raymond (1892 – 1987) of France, was exploring the *complementarity* of light waves and quanta, and realized that the principle could be applied to particles of matter; if light could be a particle then an electron could be a wave! De Broglie conjectured that electron waves 'in orbit' around the nucleus had to fit in a whole number of wavelengths; wavelengths longer or shorter would not fit. This appealed to the professor of physics in Zurich, Erwin Schrödinger who calculated the energy levels; in 1926 he found that these matched atomic spectral lines and that electrons exhibited wave-like patterns of behaviour. De Broglie and Schrödinger had established an alternative theory to Heisenberg's, which became known as ***wave mechanics.***

A research fellow at the Cavendish Laboratory in Cambridge, Paul Adrien Maurice Dirac (1902 – 84) was born and educated in Bristol where he graduated in engineering and mathematics, then moved to Cambridge to work with Rutherford. Dirac combined the hypotheses of de Broglie, Schrödinger and Heisenberg with Einstein's special relativity theory, to sort out the quantum electron puzzle. In 1928 Dirac showed, with mathematical brilliance that electrons within the atom spin around their own axes and that Heisenberg's *quantum mechanics* and Schrödinger's *wave mechanics* were mathematically equivalent, which affirmed the notion of ***wave–particle duality***. In 1932 Paul Dirac became Lucasian professor of mathematics at Cambridge University.

THE ATOM BOMBS

In 1932 John Cockcroft a physicist at the Cavendish Laboratory split the nuclei of lithium atoms with accelerated protons, then walked around the streets of Cambridge telling everyone he met "*We've split the atom*!" At that time none of the great physicists; Rutherford, Einstein or Bohr expected that there was any practical method of releasing nuclear energy, but then Enrico Fermi in Rome bombarded uranium with neutrons and found that an unexpectedly large amount of radiation was emitted. Fermi had let the nuclear genie out of its bottle. By 1939 Niels Bohr was able to give reasons why the nuclei of uranium-235 were fissile; that they could disintegrate, emit neutrons and if sufficient uranium-235 was available could start a chain reaction; a series of rapidly multiplying nuclear fissions, which would continue until all the fissionable material was consumed.

Then an artificial element plutonium-239 was found that could also give a nuclear chain reaction capable of making an atomic explosion. In 1939 at the start of the war in Europe the fear in Britain and the United States was that Hitler would develop an atom bomb to win the war. The Manhattan atomic bomb project, which employed over a hundred thousand people, was started. Robert Oppenheimer administered the project, which included Enrico Fermi, Edward Teller, Leo Szilard and many other scientific refugees from Nazi Europe.

In 1943 Niels Bohr was smuggled from Copenhagen to England. He was very miserable about the enormous dangers of nuclear weapons; moreover he had always been a pacifist. When he arrived in London, Bohr contrived a meeting with Winston Churchill to persuade the statesman of the need for a universal agreement never to use nuclear weapons, which would involve letting the Russians know about the technology. Bohr thought this could introduce an era of lasting peace. Churchill, who was in the middle of planning the invasion of Normandy, was unimpressed by these arguments and after Bohr left for America, recommended to President Roosevelt that the physicist should be arrested as a traitor. Roosevelt gave Bohr a more sympathetic interview and probably promised that the United States would not be the first to use the atom bomb.

On July 16th 1945 the first atom bomb was exploded at a testing site in New Mexico then three weeks later on August 6th a uranium-235 atom bomb, equivalent to 20,000 tons of TNT was exploded over the Japanese city of Hiroshima. It destroyed two thirds of all structures over an area of four square miles in the heart of a city with 343,000 inhabitants, immediately killing 66,000 and severely injuring 69,000. Three days later a plutonium-239 atom bomb, equivalent to 21,000 tons of TNT, was exploded over the smaller city of Nagasaki killing 39,000 and injuring another 25,000 people. The explosions produced very high temperatures and fireballs, which created convection currents that sucked dust up into the fireballs and created enormous mushroom clouds.

WHAT IS LIFE?

Best known for his whimsical thought experiment on a cat, to demonstrate quantum uncertainty, Erwin Schrödinger (1887 – 1961) was a remarkable physicist and philosopher with a strong interest in biology. His father was a prosperous Viennese lino manufacturer who had devoted himself to botany and encouraged his son to read *The Origin of Species* when he was still at school. When Mendel's work on inheritance was rediscovered in 1900 by de Vries who put forward his own theory of mutations, Erwin Schrodinger became interested in biological inheritance, chromosomes and gene theory. At this time *'genes'* were even more hypothetical than *'atoms'*.

Nonetheless he chose to study mathematics and physics at the University of Vienna where he came to understand Boltzmann's notion of statistical probability in thermodynamics. So when Schrödinger took the chair of physics at Zurich, he worked on problems related to the statistical theory of heat and was convinced *"that the laws of chemistry and physics were statistical throughout"*, which countered the determinism of classical physics.

The thought-experiment of a cat in a box, with a single radioactive atom, which might or might not disintegrate to trigger the release of felinocidal cyanide gas, characterized the indeterminacy principle. However much is known about the atom before it is placed in the box, you cannot determine when or if it will disintegrate or if the cat will be alive or not; both possibilities exist until the box is opened.

In 1927 Schrödinger succeeded Max Planck in the chair of theoretical physics in Berlin but became outraged at the actions of the Nazi regime in 1933 and moved to Oxford where a few days after his arrival he learned that he had been awarded the Nobel Prize for Physics jointly with Dirac. Schrödinger had a complex family life and at this time had a wife and a mistress with whom he had a daughter. He felt uncomfortable in Oxford and so went instead to Graz in his native Austria. After the Nazi annexation of Austria, Graz University was renamed Hitler University and Schrödinger was sacked for 'political unreliability'.

In September 1939, Schrödinger as a foreign émigré became an enemy alien in Britain, so he accepted the patronage of Eamon de Valera, the President of Ireland, President of the League of Nations and past professor of mathematics at the University of Dublin; and joined the Institute for Advanced Studies in Dublin for the duration.

It is alleged that he had another two daughters with Irish mothers and it maybe that his children, being of varied nationalities, renewed Schrödinger's interest in genetics. In Dublin he thought about the meaning of life, and in particular about heredity, gene structure and genetic mutations, in the light of statistical physics. Schrödinger addressed a serious question, which involved both physics and biology, in his book *What is Life?*, which he wrote in 1944. He also collaborated with the British geneticist JBS Haldane (1892 – 1964) of University College, London, who wrote a book with same title published in 1947.

In Schrödinger's view the dilemma lay in relation to the size of a gene molecule. His initial assessment, based on electron microscope evidence, was that it might be as large as a million atoms; however alternative deductions suggested the number might be only in the thousands. Reviewing gene structure in the light of statistical probability and quantum mechanics he wrote -

"How can we, from the point of view of statistical physics, reconcile the facts that gene structure seems to involve only a comparatively small number of atoms but nevertheless displays most regular and lawful activity with a permanence that borders on the miraculous?"

Schrödinger quoted the example of the gene for the 'Habsburg lip', which was evident in portraits and photographs of members of the Habsburg dynasty from the sixteenth to the nineteenth centuries. The abnormal gene for the deformity had been carried and replicated faithfully from generation to generation, unperturbed by thermodynamic entropy despite remaining at temperatures of about 37°C for centuries. Schrödinger conjectured that the stability of the molecular structure of a gene can be explained by quantum theory and that the *'quantum jump'* from one molecular configuration to another requires more energy than is normally available.

The conjecture explains why mutations are usually rare and why mutation rates increase when plants or animals are irradiated with high-energy x-rays, which provide enough photon energy for a *quantum jump*. He postulated on theoretical grounds that genetic molecules must be crystalline to retain stability but that the crystals would have to be irregular, or *aperiodic* to carry their complex genetic information.

Like most scientists Schrödinger assumed that genes would be proteins because chromosomes contain proteins, but shortly after his book was published, a bacteriologist at the Rockefeller Institute in New York showed that hereditary traits could be passed from one bacterial cell to another by purified molecules of ***deoxyribose nucleic acid* (DNA).** This was known to exist in the chromosomes of all cells and a few researchers thought that DNA rather than proteins might be the key to finding out how genes determined inherited characteristics.

During the Second World War, Maurice Wilkins (1916 – 2004) had worked in California on the Manhattan atomic bomb project. After Hiroshima he become disillusioned with nuclear physics and changed to biophysics instead. In 1946 he joined the Medical Research Council centre at King's College in London and began to study DNA crystals using x-ray diffraction. The x-ray photographs were complex and difficult to interpret so the centre appointed an expert x-ray crystallographer, Rosalind Franklin (1920 – 1958) to assist him.

At the Cavendish Laboratory in Cambridge in 1951, where Sir Lawrence Bragg was the director there were two junior researchers who had both been inspired by Schrödinger's book and its arguments. The oldest, Francis Crick (1916 - 2004) was a physicist and James Watson (1928-) was an American travelling scholar in viral genetics. Crick and Watson attempted some unauthorized DNA studies and tried to build molecular models. The first model was such an abject failure that Bragg commanded Crick to get on with his PhD, told Watson to study the tobacco mosaic virus and instructed them to leave DNA crystallography to Wilkins and Franklin.

It so happens that the tobacco mosaic virus contains significant amounts of ***ribose nucleic acid (RNA)***. Using a powerful new x-ray tube at the Cavendish, Watson showed that RNA had a helical molecular structure with a single strand, which he was able to model. He came to realize that DNA was the template on which RNA chains were made. In turn, RNA chains were likely candidates for protein synthesis. With further x-ray information from King's College, Watson and Crick intensified their efforts and at last produced a successful double helix model of the DNA molecule [9]. They published a paper in the April 1953 edition of *Nature* entitled

A Structure for Deoxyribose Nucleic Acid and also made a public announcement in the bar of the Eagle, the nearest public house to the Cavendish that they had discovered the *'Key to Life'*. This was appropriate, because according to Watson in his autobiographical account, *The Double Helix* published in 1968, Crick seems to have spent almost as much time in the pub as in the lab.

In 1962, Watson, Crick and Wilkins, a geneticist and two physicists, were awarded the Nobel Prize for Medicine or Physiology for discovering the structure and role of *Deoxyribose Nucleic Acid*.

Watson and Crick were right to claim that DNA was the *'Key to Life'* rather than life itself, because dry DNA crystals outside a living cell are inert, though they still fulfil the requirements of an *aperiodic crystal*. It is only in the nuclei of living cells that DNA implements its coded instructions for cell maintenance and replication.

In 1962, Watson, Crick and Wilkins, a geneticist and two physicists, were awarded the Nobel Prize for Physiology or Medicine for discovering the structure and role of DNA.

The prize is not awarded posthumously and so Rosalind Franklin's crucial role in the discovery could not be recognized. As a crystallographer she had worked with high energy x-rays for nearly ten years, probably without adequate radiation protection.

It seems possible that Rosalin Franklin's death from ovarian cancer at the age of 37 years was caused by excessive exposure to ionising radiation.

Genes are contained in large numbers, attached end to end in long double stranded helical molecules of nuclear DNA. There are 23 pairs of chromosomes in the nuclei of nearly all the cells in the body; each chromosome contains a DNA molecule, which if extended would measure several centimetres in length but which is bound to chromosomal protein. There are 20,000 to 25,000 genes in human DNA in each cell and approximately 3 billion nucleotide base pairs. Chromosomal DNA is like the musical score for a very, very large orchestra playing numerous inter-related symphonies at any one time.

Genes control cell division and inheritance from parents to children. They also control all the functions of a cell by making RNA, which spreads from the nucleus throughout the cell and controls the formation of specific functional proteins. Schrödinger was right about genes; DNA is highly durable and its replication over innumerable divisions is extremely reliable. Very small amounts left at a crime-scene or an archaeological site can provide identification years after an event.

A consequence of defining the genomes of microbes, plants and animals is the realization that there are large amounts of genetic DNA code common to all living organisms. We humans share about 98.8 per cent of our DNA with chimpanzees.

Bacteria are able to exchange 'DNA cassettes' which can modify behaviours such as antibiotic resistance; viral DNA can insinuate itself into the genes of plants and animals. Further DNA research is likely to promote a re-evaluation of current evolutionary theory. Schrödinger wrote about the role of mutations in evolutionary development and the possibility that *'quantum jumps'* in the chemical configuration of a gene in a parental gamete (male sperm or female egg cell), would alter the structure and behaviour of the offspring in a significant way, quite unlike the gradualism proposed by Darwin in his evolutionary theory.

Richard Dawkins, an evolutionary biologist, wrote in his book *The Blind Watchmaker* about what he called a *macromutation*, which occurs in mutational experiments on fruit flies, which exhibit *'antennapaedia'* in which the normal feelers of the species are replaced by another pair of legs. This mutation, usually induced by x-rays, results from a mis-transcription of normal DNA, and clearly does not improve the fruit fly mutant's survival prospects. However Dawkins noted that modern snakes have far more vertebrae and body segments than their ancestors. He postulated that new identical snake segments could be added by *'genetic instructions to a developing embryo'* resulting from a single mutational step; in Schrödinger's terms a *'quantum jump'* from one stable molecular configuration to another. Mutant DNA genes breed as true as the originals before them. From a snake's point of view there is probably significant evolutionary survival advantage in having more body segments than your ancestors; according to Professor Dawkins the fossil record suggests so.

RESUMÉ

The cardinal discoveries in physics, made in the ten years around 1900, ignited a scientific revolution, which has transformed the understanding of matter and radiation, of chemistry and biology, in ways unimaginable to scientists of that time. In retrospect the developments can be divided into two inter-related strands; nuclear physics and quantum electrodynamics.

The *radioactivity* of *uranium* was discovered in 1896 and followed shortly by the discovery of other *radioactive elements; thorium, polonium* and *radium* with the associated emission of *alpha* and *beta radiation*. The transmutation of radioactive elements into different elements, particularly the decay of radium and its daughters, was demonstrated by 1905. Shortly after, the accepted *atomic theory* was that the mass of an atom is concentrated in a small central *nucleus* composed of *positively charged protons* and of *neutrons*, which together account for the *atomic weight* of an atom. The nucleus is surrounded by *negatively charged electrons* in shells or orbitals, equal in *atomic number* to the *protons* in the *nucleus*.

Radioactive atoms have unstable nuclei that decay emitting high-energy *alpha* and *beta* particles, and also *gamma* radiation. In 1932, lithium atoms were split with accelerated protons. Then in 1940 a controlled neutron chain reaction was started in a nuclear reactor or pile, where atoms of uranium-235 were bombarded with neutrons in preparation for the manufacture of military atomic bombs. Two of these were exploded over the Japanese cities of Hiroshima and Nagasaki in 1945.

The shock and awe and horror of these bombings have damaged respect for nuclear physics. Although the peaceful use of nuclear energy may be an effective means to counter '*global warming*', many are steadfastly opposed to it. The hope of harnessing the fusion of hydrogen nuclei to form helium with the production of cheap nuclear energy has yet to be realized.

Photon-electron physics is much less doom-laden than *nuclear physics*. The discovery of *x-rays* (1895), *electrons* (1897), the *photo-electric effect* (1899) and light *quanta* led in 1905 to a *quantum theory of light* and its interaction with *electrons*. The link between the *quantum energy* of light emitted or absorbed by a hydrogen atom and the *energy levels* of its single electron then led to theories of *quantum mechanics*, which gave coherent mathematical explanations applicable to all atoms. The quantum model of atoms with *electron shells* revealed the mechanisms of *chemical bonding* in inorganic and organic molecules. The discoveries of *quantum electro-dynamics* have fuelled a revolution in electronics and computing.

Sixteen years after their discovery, *x-rays* were shown by crystal diffraction to be high-energy *quanta* and then x-ray crystallography was developed as a precise tool to analyse molecular structures.

In 1944 the physical nature of living organisms, particularly of their chromosomal molecules, was reviewed in the light of quantum theory and the laws of thermodynamics. This analysis conjectured that the key genetic molecule for living organisms must be an *aperiodic crystal*.

In 1953 a geneticist, two physicists and a physical chemist, using x-ray crystallography, demonstrated the complex molecular structure of genetic *deoxyribose nucleic acid*, which provides coded instructions for the functions of living cells. This understanding of the *'Key to Life'* came from the laws of quantum physics and technologies that arose from them.

This biological revolution continues. The Human Genome Project was started in1990 as an international programme and by 2003 had mapped the entire human genetic code or codes. It optimistically promises to have a major impact on personal *genomics*, where everyone will know their own genetic code, which will indicate disease risks and behaviour patterns. Already many mutations or faulty genes associated with an increased risk of cancer (*oncogenes*) have been identified. Some sequences that indicate an increased risk of degenerative brain disease or autoimmune disorders have also been revealed by the Human Genome Project in 2007.

The hope is that treatments will be tailored to our individual needs and side effects will be rare. *Gene therapy* to replace or correct faulty sequences (*DNA spelling mistakes*), may become possible. These future advances will be welcome but there is the threat of a return to a form of mechanistic determinism as intense as that of the 17th century and uneasiness about *'meddling with nature.'*

The discovery of the structure of DNA and its role in regulating the functions and structures of all the cells in the body, as well as genetic inheritance, is further evidence of the close relationship between physics and human biology.

NOTES

1. Until 1957 the unit of x-ray intensity was the ***roentgen,*** which was the amount of radiation that produced a unit of electric charge per unit mass of air. It is no longer used because it only applies to air and has been replaced by the SI unit, the ***gray (Gy),*** which is the dose of ionising radiation, which delivers ***one joule of energy per kilogram of recipient mass.***

2. Alexander Litvinenko, a former lieutenant colonel in the Russian secret police and outspoken critic of the Russian government, who found refuge in Britain, was murdered by poisoning with polonium-201 and died in a London hospital. The main suspect was Andrei Lugovoi, a Russian secret agent who returned to Moscow, leaving an incriminating trail of polonium 201 as he went. The Russian government refused his extradition for trial in the UK.

3. The ***curie (Ci)*** was the unit of radioactivity and represented the quantity emitted by one gram of radium bromide; however it is no longer used officially and has been replaced by the SI unit, the ***bequerel (Bq),*** which represents ***one disintegration*** *of an atomic nucleus* ***per second.***

4. In all there are nine transmutations from radium to stable lead with half-lives ranging from 3 minutes for Radium A, to 1602 years for Radium-226. Radon gas with a half-life of 4 days remained an '*emanation*'.

The paper also described three types of radiation associated with the transmutations, which Rutherford identified as alpha, beta and gamma – α, β and γ rays. Only the transmutations of radium C and radium E were associated with β and γ rays; the transmutations of radium, radon, radium A and F only gave α rays.

Radium	Radon gas	Radium A	Radium B	Radium C	Radium D	Radium E	Radium F	Lead
Ra-226	**Rn-222**	**Po-218**	**Pb-214**	**Po-214**	**Pb-210**	**Bi-210**	**Po-210**	**Pb-207**
1602 years	4 days	3 mins	21 mins	&**Tl-210**	40 years	6 days	138 days	stable

The chain of radioactive decay (linked transmutations) from radium-226 to stable lead [with radioactive half-lives].

5. **A gram-atom** is the quantity of a chemical element whose mass in grams numerically equals its atomic weight. The number of hydrogen atoms in one gram of the gas or of carbon atoms in a12 gram diamond is 6×10^{23}! This is a ginormous number and a million times more than all the seconds of time since the Big Bang, nearly 14 billion years ago.

6. If you go to the theatre or a match in Italy you ask *'Quanto?'* which means *'How much?'* A standard seat might cost fifty euro (**€ 50**); no more and no less. With only € 49 you won't get in and € 51 is too much. A luxury seat might cost **€ 100**; no more and no less. It costs exactly €50 to change from a standard seat to a luxury seat.

In the same way a photon absorbed into an atom must have exactly the right *quantum* of energy (***hv***) to lift an electron to a higher energy level within the atom. And a photon with the same *quantum* energy (***hv***) or a light wave of exactly the same frequency (*v*) is emitted when an electron falls from the high energy level back to its original level. Photons with *quantum* energies that don't match electron energy levels will not be absorbed (allowed in) and can't be emitted. Absorption or emission of *quanta* with specific frequencies accounts for characteristic bands for different elements seen in the spectroscope.

7. The atomic number [Z] of lead is 82 so all the isotopes of lead have 82 protons in their nuclei, but radioactive lead-214 has 132 neutrons and radioactive lead-210 has 128 neutrons, whereas stable lead-207 has only 125 neutrons

8. Anatomical elements.

element	% body weight	atomic number [Z] protons/electrons	atomic weight protons+neutrons
hydrogen H	9.5	1	1.0
carbon C	18.5	6	12.0
nitrogen N	3.3	7	14.0
oxygen O	65.0	8	16.0
sodium Na	0.15	11	23.0
phosphorus P	1.0	15	31.0
chlorine Cl	0.2	17	35.5
potassium K	0.4	19	39.1
calcium Ca	1.5	20	40.1

Best & Taylor 1954

9. The basic building blocks of **DNA** are **nucleotide** molecules each composed of a **nitrogenous base,** plus **phosphoric acid** combined with **deoxyribose** (a sugar) of which the two outer helical strands of DNA are made. The rungs connecting the two **helical sugar-phosphate strands** are pairs of **bases;** either **adenine-thymine (A-T or T-A)** or **cytosine-guanine (C-G or G-C).** The order of these pairs in the DNA molecule contains the coded information of genes.

The basic building blocks of **DNA** are **four nucleotides** each made of phosphoric acid **P,** a sugar called deoxyribose **D** and either one of four bases – adenine **A,** thymine **T,** guanine **G,** cytosine **C.**

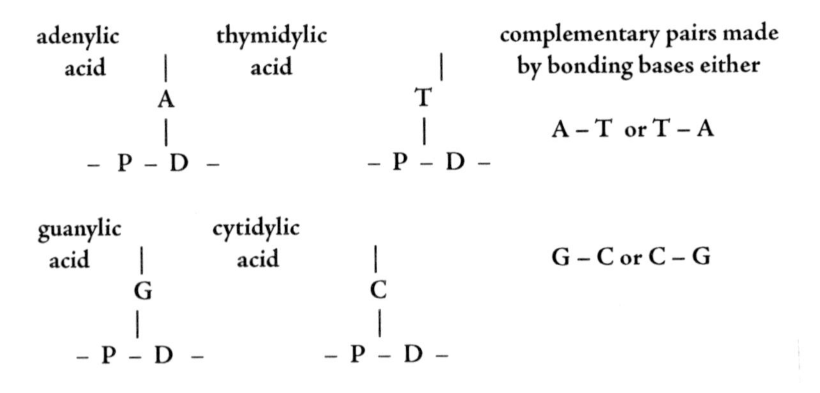

DNA is formed by **bonding of nucleotides.** Bonds between **P** and **D** form two outer strands twisted into a double helix, held together by rungs of paired bases, either **A – T, T – A, G –C** or **C – G.**

```
– P – D – P – D – P – D – P   D – P – D – P – D –
  |   |   |   |   |   |   |   |   |   |   |   |  line of
  G   C   A   T   A   C   T   G   A   C   C   T  separation
--|---|---|---|---|---|---|---|---|---|---|---|--of base --
  C   G   T   A   T   G   A   C   T   G   G   A   pairs
  |   |   |   |   |   |   |   |   |   |   |   |
– P – D – P – D – P – D – P – D – P – D – P – D –
```

The DNA molecule replicates itself by unzipping down its centre by separating the base pairs. Each separated strand forms a template for the formation of a new matching strand of nucleotides, exactly replicating the original DNA molecule. Alternatively it can form a template for protein synthesis by transcribing the sequences to make messenger RNA, which leaves the nucleus to collect amino acids, link them together and build them into proteins

RADIOLOGY – EXPLORING INNER SPACE

During the 20th century *investigative medicine* had to match the increasing sophistication of medical treatments of all kinds; to make the initial *diagnosis* for a patient of a specific disease and define its extent; monitor the response of the disease to treatment and detect complications or side effects. Accurate diagnosis is essential for effective treatment. Nothing can be worse than giving a patient treatment, such as drastic surgery, for the wrong disease. Clinical pathology has made enormous advances based on those of molecular biology, immunology and microbiology. Advances in ***diagnostic radiology***, alternatively known as ***medical imaging***, have depended mainly on technologies arising from physics. *Radio diagnosis* started with *x-radiography* then developed to include *ultrasound, gamma imaging, computer-assisted tomography (CT) and magnetic resonance imaging (MRI)*. Investigative *clinical pathology* and *radiology* have greatly improved the accuracy of diagnoses over the last hundred years and as a consequence, they have improved the efficacy of medical treatments.

The findings of *clinical radiology* cannot stand alone in making a diagnosis but have to be integrated with the patient's clinical history, medical examination and pathology laboratory results, so as to achieve a diagnosis. Otherwise there is a danger of treating a scan and not the patient.

X-RAY RADIOGRAPHY

The basic principles of *diagnostic radiography,* also known as *roentgenography,* have not changed since the discovery of x-rays and the invention of radiographic imaging by Röntgen in 1895. The method depends upon the differential absorption of x-rays as they pass through the body. A beam of x-ray photons emitted from an x-ray tube is attenuated or thinned, as it passes through different types of body tissue, to produce a projection image that is recorded on photographic film contained in a light tight cassette. When developed and fixed, the photographic negative is an x-ray radiograph (XR); the result is a projection of the overlapping shadows of three-dimensional structures onto a two dimensional image. On a standard chest radiograph the X-ray shadows of the breasts, ribs and muscles of the chest wall overly those of the heart and lungs.

Instead of x-ray photography, parts of the body can be viewed using an x-ray fluorescent screen or *fluoroscope* to see a projection image produced by x-rays transmitted through the tissues. This has the advantage of observing movement of body organs as they happen, but gives poorer image quality than x-ray photography.

Röntgen had shown that metals like lead with high atomic numbers (the atomic number [Z] of lead is 82), absorbed x-rays and attenuated an x-ray beam much more than lighter metals such as aluminium [Z=13]. When x-rays came to be regarded as photons, it was clear that the interaction of x-rays with atoms involved *photon-electron* interactions and **not** with *atomic nuclei*. A common misconception is that radiography or radiotherapy can make patients radioactive; but medical x-rays do not interact with atomic nuclei and so they cannot change stable atoms into radioactive atoms.

One of three things can happen to photons in an x-ray beam as they pass through a volume of matter. They may:

1. Pass straight through without hitting any atoms – **transmitted radiation**: or
2. Be deflected on contact with electrons to produce – **scattered radiation**: or
3. Hit electrons with enough quantum energy to eject those electrons from atoms and form ions – **ionizing radiation.**

Photons are more likely to collide with, and be absorbed by or deflected by atoms with larger numbers of electrons and greater atomic numbers, than those with fewer electrons. Compact bone has a high effective atomic number [Z=13] because it contains calcium [Z=20] plus phosphorus [Z=15], with a high electron density and so attenuates an x-ray beam (absorbs or deflects x-ray photons) more than water [Z=7.5] or fat [Z=6]. Tissues with higher **electron densities** stop more photons.

Bone and metal fragments stop most x-ray photons in the beam; few reach the film in their shadow, which remains clear when developed. Air in the lung and gas in the bowel have low electron densities, absorb few photons and so they look black on the film. Tissues of water density, like muscle, tendons, cartilage, liver, heart and blood vessels show as intermediate shades of grey; fatty tissue is less dense than water and so looks slightly darker than muscle on radiographs.

With fluoroscopy these values are reversed because photons that pass easily through the lungs make them seem brighter on the fluorescent screen than the heart and chest wall muscles. Bones and metal fragments look black. In photographic terms, radiographs are *negative images* and fluoroscopy gives *positive images*. Both provide ***two-dimensional projections of the varying electron density of three-dimensional anatomy.*** On a chest radiograph, shadow images of the ribs, spine, heart, lungs and soft tissues overlap. Both radiographs and fluoroscopic images represent the passage of photons through different types of body tissue and can be regarded as methods of x-ray **transmission imaging.**

Röntgen established the basic principles of radiography in 1895, but the equipment used over the next twenty years was erratic and dangerous, because of ionising radiation; also from poor electrical safety with the risk of electrocuting radiographers or patients. Early x-ray tubes contained some gas and emitted variable amounts of x-rays of different energies. Their beams contained too many, low energy 'soft rays', which could not penetrate much deeper than the skin; so were useless for imaging, yet could cause burns. Transmission imaging needed high-energy 'hard rays'. Compared with modern tubes the output of x-rays was poor; exposures took minutes to blacken film. Scattered x-rays blurred pictures, so radiography could only show fractures, metal fragments and large shadows in the lungs. Fluoroscopy was easier but caused dermatitis of the physician's face and hands; image quality was less than for radiographs and it could miss significant fractures or lung lesions.

William Coolidge (1873 – 1975) studied physics in Germany where he obtained his PhD before returning to America. In 1913 whilst working for the General Electric company in Schenectady, New York, he designed a completely evacuated x-ray tube with a tungsten anode that produced a more focused and stronger beam of x-rays.

After the First World War Coolidge tubes became available in Europe and America and significantly improved the quality and range of radiography. Most modern x-ray tubes are based on his original design. In an x-ray tube photons are created at the anode when high-energy electrons from the cathode bombard it. Energy is given to the electrons by a high voltage applied across the tube between the cathode and the anode. A transformer supplies voltages of between 40,000 and 120,000 volts (40 - 120 kV) to give an x-ray beam enough 'hardness' for diagnostic imaging.

A small current, measured in milliamps (mA), heats the cathode and controls the amount of electrons during each exposure and therefore the quantity or 'dose' of x-ray photons produced at the anode. Apart from a small aperture for the primary x-ray beam, the tube is enclosed in a protective metal casing. Aluminium filters placed at the tube aperture absorb most of the useless 'soft rays' from the beam and reduce unnecessary irradiation. A diagnostic x-ray machine allows a radiographer to direct the primary beam at the correct portion of anatomy.

Radiographic film, like ordinary photographic film is much more sensitive to visible light than x-rays. If the film is sandwiched between **intensifying screens,** which fluoresce when exposed to x-rays within a lightproof cassette, the amount of radiation needed for the same degree of film blackening is reduced to one tenth. By the 1920s the use of more sensitive film, intensifying screens and better x-ray equipment helped reduce radiation doses and exposure times to seconds rather than minutes. It became easy and safe to look inside a human body without needing to commit surgery.

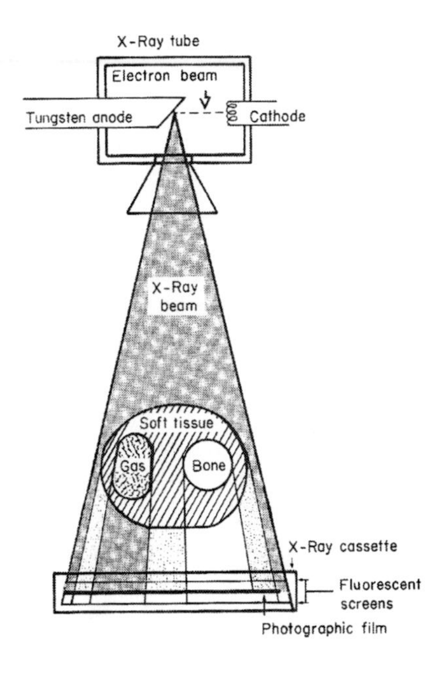

32. Conventional X-ray Radiography

PLAIN RADIOGRAPHY

Plain radiography, that is, without the use of artificial contrast agents like barium, iodine media or injected air, is limited to distinguishing only four groups of tissues each with different electron densities – **gas** in the airways, lungs and gut; **water-density** muscle, tendons, cartilage, blood vessels, brain, liver and kidneys; **fat density** of adipose tissue and **calcium** in bones and stones. The plain examinations, which are the most effective, rely on radiographic contrast between air or bowel gas, water-density tissues and bone.

In the developed world, plain radiography is still the most commonly performed of all diagnostic imaging procedures but is no longer dependent on photographic film using costly silver halides. Many modern radiology departments now claim to be 'silverless'. Digital x-ray cassettes contain reusable, photosensitive phosphor sheets instead of silver halide film and require smaller doses of radiation to obtain diagnostic images. Instead of photographic development the electronic image on a phosphor sheet is 'read' by a digital laser, then stored and analysed on a computer.

The chest extends from the neck to the waist and is supported by the thoracic spine, ribs, shoulder girdles and all the muscles and sinews attached to them. It is separated from the abdomen by the muscles of the diaphragm. The chest wall and diaphragm form the boundaries of the thoracic cavity. External to the chest wall are the skin, subcutaneous tissues and the breasts. The lungs, covered by pleural membrane, lie within the thoracic cavity on either side of the central mediastinum. The mediastinum contains the heart and great blood vessels as well as the oesophagus and the trachea, which communicates with the major bronchi for each lung.

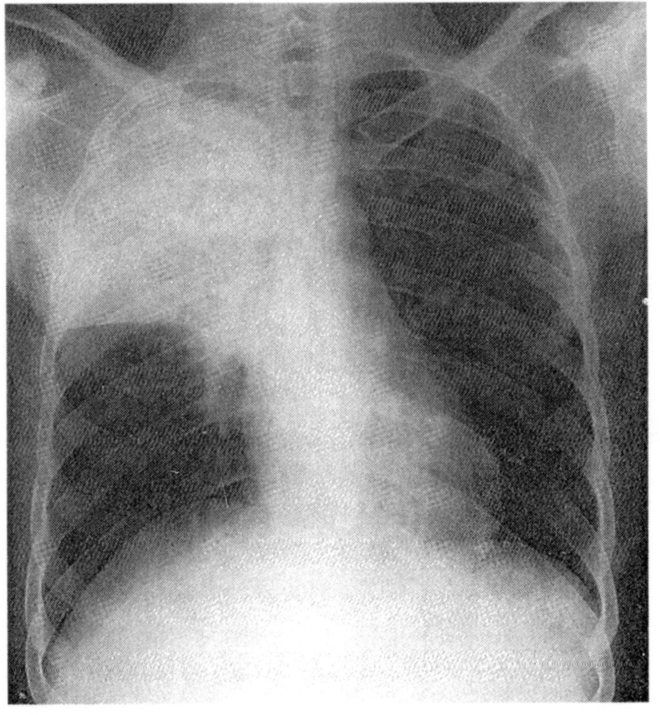

right

33. Chest radiograph of a patient with right upper lobe pneumonia. The exudates in the infected lung are dense like water and so **pneumonia** shows up as 'white' compared to the normal dark lung. A pleural fissure separates the upper lobe from the rest of the right lung and has stopped pneumonia spreading into the lower zones. The left lung is normal in appearance. Chest radiographs, like all medical images, are viewed as if looking at the patient, so that the right side of the patient is to the left side of the viewer.

On a radiograph of the chest, the shadows of all these anatomical structures are superimposed. The **chest x-ray radiograph (CXR)** benefits from radiographic contrast because on full inspiration, the air-filled lungs are outlined at their boundaries with the chest wall, with the diaphragm below and with the mediastinum lying between them. As the heart is the largest component of the mediastinum, its outline on a radiograph is sometimes called 'the cardiac silhouette' and its apparent size can reflect the true size of the heart. It may be enlarged in heart failure or hypertension.

The pulmonary blood vessels within the lung are outlined by air in pulmonary alveoli to give a picture of the major arteries and veins within the lung fields. Diseases such as pneumonia, tuberculosis and tumours are also of water-density and are visible on the CXR in contrast to the air in normal lung tissue surrounding them.

Tuberculosis is one of the oldest human diseases. Known as *consumption* or *phthisis* it was called the 'Captain of all these Men of Death' by John Bunyan and accounted for a fifth of the deaths in London's Bills of Mortality in 1651. After Koch's discovery of the tubercle bacillus in 1882 and the realization that the disease was infectious, isolation of patients in sanatoria to control its spread became a generally accepted policy, which was partially effective. Tuberculosis had such a devastating impact because it was a chronic disabling disease that affected children and young adults, especially in the crowded cities of Europe and Asia where it was the commonest cause of death between15 and 24 years of age.

Although tuberculosis was common, its pathological progress was not understood before chest radiography became available. The deflated lungs of a corpse are difficult to examine and autopsies of TB patients only showed the severe scarring and cavitations of terminal disease. How the infection started and developed were unknown. Then in the 1920s, for the first time, early tuberculosis could be identified in living patients and disease progress monitored with chest radiography.

Other lung diseases such as chronic abscesses and growths, which had masqueraded as *consumption* could be diagnosed separately. The characteristic radiographic changes of pneumoconiosis (pulmonary dust disease) such as silicosis and 'coal-miners lung' revealed the hazards of occupational dust exposure.

Tuberculosis in its early stages causes only trivial symptoms and yet the patient may spread infection within the family and in the workplace.

In 1938 mass-screening chest radiography was introduced to identify those with early disease within large groups of individuals, such as nurses, students, soldiers and factory workers, to treat them at an early stage with some hope of cure and to isolate them, so as to prevent the spread of the disease. Approximately four previously unsuspected cases of active disease were discovered per thousand examinations. After World War II mass chest radiography became an established method of preventive medicine in developed countries and caused a reduction in disease prevalence even before effective anti-tuberculous drugs became available in the late 1950s. After this the numbers of patients with TB fell to an all-time low in the early 1980s; sanatoria closed and mass-screening chest radiography was no longer needed in Europe or America; however the disease persisted in sub-Saharan Africa and Asia.

When the human immunodeficiency virus (HIV) erupted in the 1980s it made patients more vulnerable to opportunistic infections like TB and it has now returned to the crowded cities of Europe and North America. In London there were 3,834 new cases of TB in 2005; some of its boroughs had more than 70 new cases per 100,000 people in that year; less than 5 percent were co-infected with HIV. It is conceivable that mass-screening chest radiography for TB may return.

The chest XR has diagnostic limitations. Detection of solitary lung tumours, less than half a centimetre in diameter, is erratic because overlying ribs can obscure such small shadows; even large bronchial cancers lying within the mediastinum are also invisible. Obstructive pulmonary disease only produces CXR changes when it is advanced and large pulmonary emboli blocking blood flow to the lungs usually produce no change in appearance. In spite of these limitations the CXR continues to be used extensively in modern hospitals and clinics; other imaging techniques have been devised to cover its diagnostic defects.

Mammography: radiographic examination of the female breast is relatively recent. Although x-ray diagnosis of breast disease was first reported in Strasbourg in1951 mammography was not generally available until 1960, when Dr Robert Egan from the MD Anderson Hospital in Texas introduced a much-improved technique. Mammography is different from other examinations because the radiographic contrast; the difference in electron density, between the fatty [Z=6] and glandular tissues of water density [Z=7.5] within the breast is very slight. It so happens that the contrast is best with 'soft' x-rays of only 26 – 30 kV and special x-ray tubes are required.

In young women, especially if lactating, the breasts are very glandular so contrast is poor; however after the menopause there is more fatty replacement of glandular tissue so that small impalpable cancers, which are of water density, are more easily perceived. Another feature of breast cancers is that about 50 percent have fine micro calcifications evident on high-resolution film; however other benign breast diseases also show micro calcifications so that x-ray **guided needle biopsy**, to obtain tissue for microscopic diagnosis is required.

By 1962 Egan had diagnosed fifty impalpable breast cancers with mammography. A mass-screening programme, similar to that for pulmonary TB, was organized in New York City to see if early diagnosis from mammograms could reduce the mortality from breast cancer. 30,000 women aged 40-64 years were examined between 1963 and 1970. The results were positive with an overall mortality reduction after 18 years of about 25 percent. Women over 50 years benefited more than younger women, probably because cancers are easier to identify in more fatty postmenopausal breasts. Similar screening programmes have been established elsewhere in the World.

Mass breast screening is not preventative; its aim is to find lesions earlier, so that treatment is more likely to be effective.

Women who undergo regular breast screening have to be aware that rapidly growing breast cancers can still present in the two or three year intervals between screening examinations.

Plain radiography of bones and joints also improved between the two world wars and was crucial to the development of new orthopaedic surgical techniques for treating fractures, dislocations, bone infections and correcting paediatric deformities. Mature compact bone contains large amounts of calcium phosphate, which gives excellent radiographic contrast with other tissues around it: such as muscle, tendon and cartilage.

Bone is formed by the ossification of cartilage in growth plates at the ends of long bones. The growing bones of children have a much larger proportion of unossified cartilage than adults and it is possible to assess a child's skeletal age by the extent of ossification of the growth centres in the small bones of the hand. Characteristic radiographic changes due to abnormal growth of the bones of infants with rickets, scurvy and hypothyroidism led to earlier diagnosis and treatment of these potentially crippling diseases, which were prevalent in the early 20th century. Bones and joints as well as lungs can be affected by tuberculosis and the most severe form affects the spine. Spinal radiography is of great value in detecting early tuberculous destruction of vertebrae and intervertebral discs, which without treatment will lead to severe kyphosis or 'hunch-back' deformity. Crippling spinal TB is still common in Africa where there is often a shortage of radiographic equipment suitable for diagnosing the disease in its early stages.

Malignant tumours can affect bones and are of two kinds: **primary** tumours which arise directly in the bones occur most often in adolescents and are fortunately rare: **secondary** tumours or metastases; from primary cancers in other organs such as the breast, prostate and lung, are more prevalent in older patients.

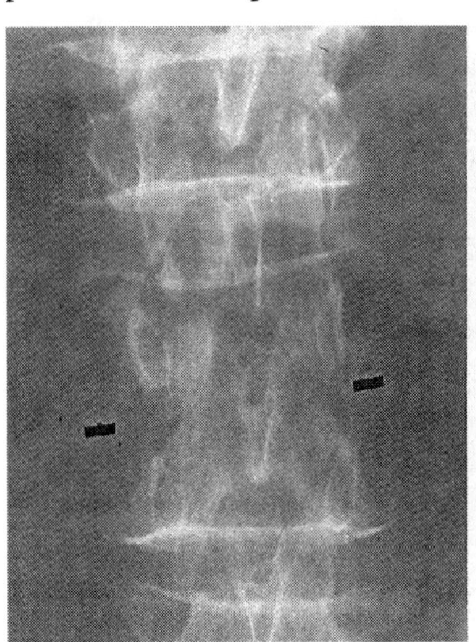

frontal view

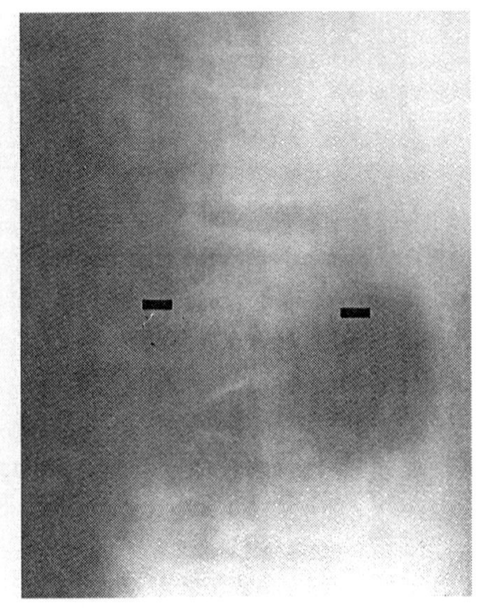

conventional midline tomogram

34. Spinal tuberculosis. The frontal and lateral tomographic views show that the **intervertebral disc (black markers)** between two lumbar vertebral bodies has been **eroded** and its margins have disappeared, so that the two vertebrae are merged into one. As this destructive process spreads to other levels in the spine it leads to 'hunch back' deformity

Like the chest XR, plain bone and joint radiography has diagnostic limitations. Its prime role is the detection of fractures as soon after injury as possible, which depends on defining a radiolucent gap between bone fragments or identifying bands of impacted bone. With good radiography, detection of fractures is very accurate but undisplaced; 'hairline' fractures can be invisible. Such fractures often occur in the scaphoid bone in the wrist and in the neck of the femur, near the hip joint. If not detected and immobilized these fractures block blood supply to bone fragments. The bones die and crumble, leading to long-term disability. Another limitation is that skeletal radiographs project images mainly of compact bone.

Metastatic bone cancer usually starts in the marrow and erodes compact bone from inside; so significant bone destruction must occur before radiographic bone changes are evident. Bone infection: osteomyelitis, may be present for 10 days before XR changes can be seen.

Obstetric radiography of the pregnant abdomen was a form of skeletal radiography, which pictured a baby's skeleton *in utero*. Ossification of the fetal bones progressed through pregnancy and radiography could be used to assess maturity; evidence of fetal growth retardation or death; the number of babies present, the alignment of the baby's head and its relationship to the mother's pelvis before labour.

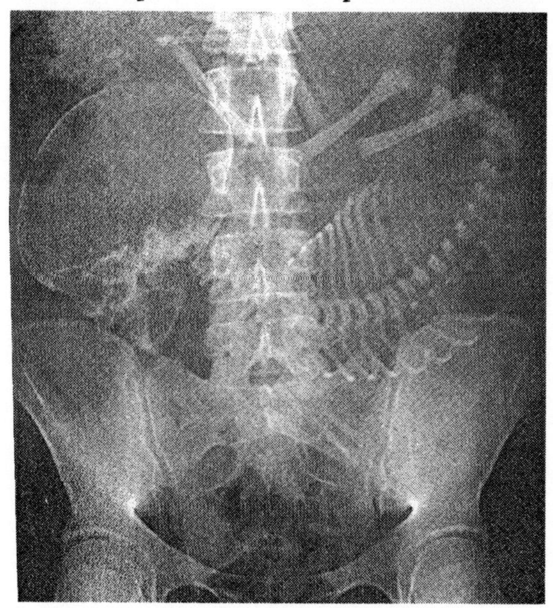

35. Abdomen radiograph in late pregnancy taken in 1970. This baby should have been positioned with its head in its mother's pelvis. It is lying **transversely** because the placenta is low in the uterus **(placenta praevia)** and is blocking the head from entering its normal position. Placenta praevia can cause catastrophic bleeding. This baby was delivered by caesarean section soon after this XR.

Up until the 1960s and before obstetric ultrasound became available, obstetric radiography was used frequently in complicated pregnancies; however it was poor at showing the placenta and abnormalities early in pregnancy, even embryonic death. It gave a big dose of x-rays to the whole baby. Obstetric radiography is now obsolete.

BARIUM

Unlike the lungs and bones, the organs of **the digestive tract**; the oesophagus, stomach, small and large bowel, do not have innate radiographic contrast. Though they may contain some air or bowel gas, this is insufficient to define their lumens (hollow passages) accurately and so artificial contrast agents are required. Only two years after Röntgen's initial discovery, Walter Cannon of Harvard University, studied the intestinal movements of cats that had been fed bismuth salts, using fluoroscopy. Barium sulphate is a stable, non-toxic compound, which replaced bismuth for medical examinations of the stomach and colon after 1910. Barium is a metal with a high atomic number [Z=56] with high electron density so that barium sulphate gives good radiographic contrast.

By 1920 the **barium swallow and meal** had become the standard method for examining the oesophagus, stomach and duodenum: to identify peptic ulcers or cancer. Soon after this the **barium enema** examination was established for diagnosing diseases of the large bowel.

Barium examinations need fluoroscopy to look at the movement of barium through the stomach and intestine to see if there is hold-up by strictures or growths and to obtain 'spot' radiographs of the organs in the best position to demonstrate disease.

Before image intensifiers became available in the 1960s, naked-eye fluoroscopic screens provided very faint and blurred images compared with radiographs; however it was possible to see the gut in motion. Barium examinations were performed in light-less rooms by dark-adapted radiologists; retinal rods at their most sensitive to the faint shadows of barium-filled gut on the fluorescent screen.

For patients drinking barium or getting an enema in total darkness, positioned by unseen hands; these examinations were not comfortable.

When visible, the clinical radiologists of those days were a scary sight. They were dressed from neck to knees in barium stained, radiation-protective lead aprons and wore red tinted goggles to preserve dark adaptation when outside their dark x-ray rooms, squinting like moles.

Image intensifiers provide significant electronic enhancement of the faint images seen on simple fluorescent screens so that intensifiers have made naked-eye fluoroscopy and red goggles obsolete since the 1970s. Barium examinations and other procedures can now be performed more quickly and safely in subdued lighting. The radiation doses for patients and staff are significantly less compared with the old method. The images from the output phosphor of the intensifier can be viewed by via a television camera, recorded on microfilm, videodisc or stored on a computer.

In 1958 a South African gastroenterologist, Basil Hirschowitz, who later became professor of medicine and biophysics in the University of Alabama, introduced the **flexible fibreoptic endoscope.** This was designed initially to examine the oesophagus, stomach and duodenum, then modified to examine the large bowel. The instrument gave direct vision of the mucus membranes, which line the surfaces of the upper digestive tract or colon, safely and with relatively little discomfort for patients. **Endoscopic biopsy** also provides pathological confirmation of malignant gastric ulcers and colon cancers. Barium studies and endoscopy continue to be complementary investigations in the diagnosis of gastrointestinal disease.

IODINE

Like the gut, **the kidneys, ureters and urinary bladder** are of water density and not normally visible on a plain radiograph of the abdomen; although calcified stones in the urinary tract, which can cause extreme pain, often are. Sodium iodide was an accepted treatment for syphilis in the 1920s and it was discovered accidentally that on radiographs it opacified the bladder. Sodium iodide was too toxic to be injected intravenously but organic iodine containing salts based on benzoic acid were introduced, which were readily excreted by the kidneys.

Most patients tolerated these water-soluble contrast media, which opacified the renal pelves and ureters that drain urine from the kidneys into the bladder. The examination was able to demonstrate obstructions caused by urinary calculi (stones) or growths and became known as **intravenous pyelography (IVP)**; later as **intravenous urography (IVU)**.

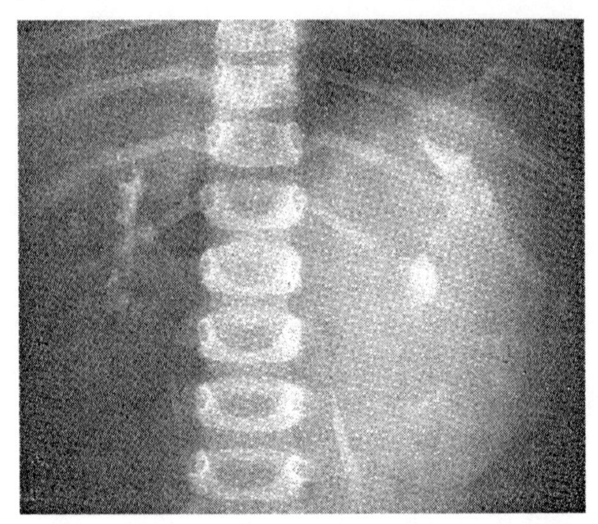

36. Intravenous Urogram after an injection of urografin in a child with a large **tumour in the left kidney**. The kidney on the right side appears normal.

Allergic reactions and other side effects were common with the early contrast media; however in the 50 years following their introduction in 1929, better compounds, which can be safely injected into veins and arteries in large amounts, were developed. These contrast media usually contain three atoms of iodine for each benzoic ring. Iodine has a high atomic number [Z=53] with high electron density and so provides radiopacity; the remaining components are not radiopaque but increase the solubility and reduce the toxicity of the contrast medium molecule. In common parlance a contrast medium is often referred to as a 'dye' but this gives a false idea because these compounds are colourless and do not stain tissues. By using larger doses of contrast medium it became possible to visualize the tissue of the kidneys themselves, as well as the ureters and bladder.

Like the kidneys and ureters, **the liver, biliary ducts, gallbladder and pancreas**, are poorly visualized on a plain radiograph. The liver is the heaviest single organ in the body and is its powerhouse; it plays a major role in carbohydrate, lipid and protein metabolism. The liver excretes bile, which contains waste products such as bilirubin; a yellow pigment formed by the breakdown of haemoglobin from old red blood cells. Bile also contains bile salts which act as detergents, crucial for digestion of dietary fats and fat-soluble vitamins; it is stored and concentrated in the gallbladder.

Acute jaundice occurs when bilirubin cannot be removed from the blood because of liver failure due to hepatitis or because the bile ducts are blocked, either by duct stones or by tumours. Increased bilirubin in the blood causes yellowing of the skin and the eyes so that the whites of the eyes turn yellow. Deciding between two causes of jaundice; hepatitis or obstruction due to a stone is vital because abdominal surgery increases the metabolic load on the sick liver and can lead to death, whereas removal of an obstructing stone can give immediate cure.

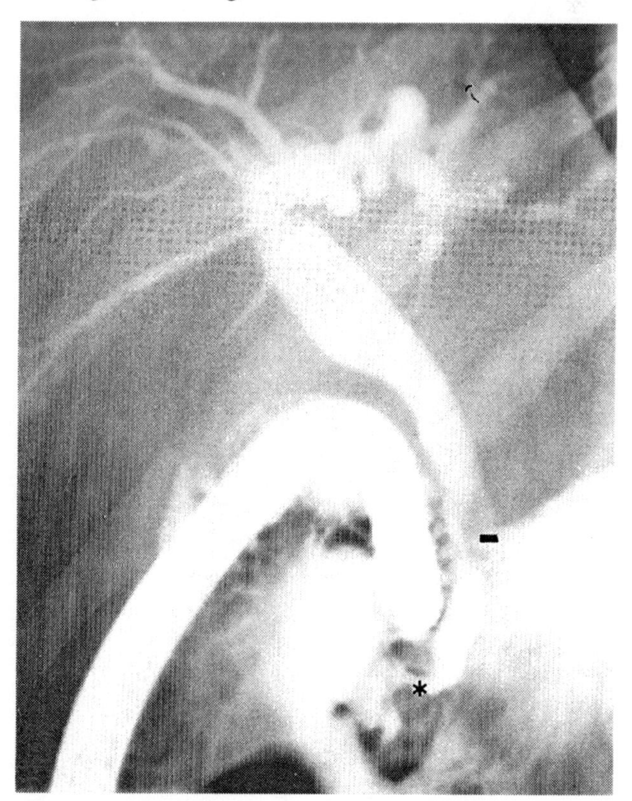

37. ERCP of a bile duct stricture. A catheter has been passed through the endoscope into the papilla of the bile duct (*) Injected contrast medium outines the bile duct to show the stricture (-).

In 1970 a Dr I Oi in Japan devised the ingenious technique of **endoscopic retrograde cholangiopancreatography** (ERCP), which combined several diagnostic technologies. Oi used a **fibre-optic endoscope** in the duodenum to pass a **fine catheter** through the scope and insert it into the papilla where the bile duct and pancreatic duct enter the duodenum. **Contrast medium** injected through the catheter then passed retrogradely, that is backward against the flow, to opacify the entire biliary tree and pancreatic ducts under **image intensifier** control and then make radiographic images. This complex test required extraordinary skill to cannulate the elusive papilla in the duodenum but the method was accurate and relatively safe for diagnosing jaundice.

By 1980 contrast media were so safe that they could be injected into the spinal canal to outline the spinal cord and show spinal tumours and prolapsed inter-vertebral discs pressing on nerves. The examination is called **myelography.**

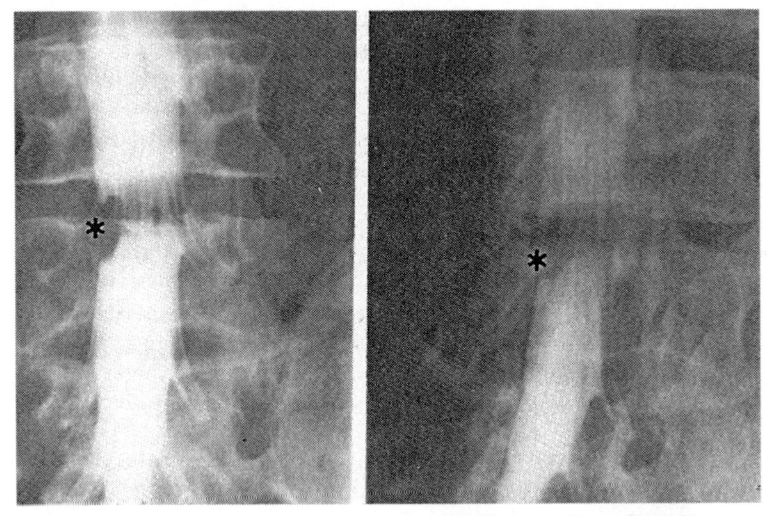

38. Myelogram of disc prolapse. Contrast medium outlines the CSF space surrounding the nerve roots in the lumbar spine and shows a disc prolapse (*) pressing on the right nerve root at the L 4 -5 level.

AIR

A plain **skull radiograph** is just that. It reveals the beautifully detailed anatomy of the bones of the skull: the cranial vault, facial bones and jaws, but does not image the **brain** within the cranial cavity. The tiny pineal gland at the centre of the brain, which Descartes thought to be the seat of the soul, is the only part that may be calcified in healthy adults. Though composed of grey and white matter and surrounded by cerebrospinal fluid, the healthy brain is of uniform radio-density on plain radiographs and therefore invisible. Some brain tumours, infections and aneurysms, can produce calcifications that may be visible on a plain radiograph, but not often enough to be of diagnostic value. Following a head injury, a fracture of the skull may indicate the severity of injury to the underlying brain, especially if fragments of bone are depressed into the cranial cavity, but the brain can be severely injured, even in the absence of a skull fracture.

In 1912 a patient presented himself to a Dr Luckett of New York City because of persistent headache for ten days following a head injury. The doctor used his new x-ray machine and showed air in the cranial cavity, which had entered through a skull fracture. Dr Luckett discovered air, which appeared darker than adjacent brain, had entered the cerebral ventricles and that air could be used as a contrast medium to outline the ventricles and surface of the brain.

In 1918 Walter E Dandy (1886 – 1946) the professor of neurosurgery at Johns Hopkins University intentionally replaced ventricular cerebrospinal fluid with air so as to obtain radiographs of the ventricles; initially by open operation, then later by injecting air into the lumbar spinal canal so that it floated upwards into the cranial cavity and ventricles. This test was known as **ventriculography** or **air encephalography,** and was not a pleasant experience for a patient because it induced severe nausea and a headache that lasted for at least several days. Ventriculography could also be dangerous, but before brain CT was invented, it was the only way of localizing some brain tumours and demonstrating surgically treatable congenital brain abnormalities. It is now obsolete.

ANGIOGRAPHY

As early as 1899 an anatomist in London had injected mercury into the arteries of a dead baby and made a radiographic map of its arterial tree. Within a few years, several anatomists in Europe,

America and Australia had made arterial maps and atlases of dead human limbs, hearts and brains, using mercury and concentrated salt solutions of toxic heavy metals. Living arteries were more difficult to examine, but a distinguished Portuguese physician showed how it could be done.

António Caetano de Abreu Freire Egas Moniz (1874 – 1955) was one of the most remarkable doctors of the 20th century. After qualifying in medicine at the University of Coimbra, the intellectual capital of Portugal, Moniz studied neurology in Paris then returned to Coimbra where he was appointed chairman of the neurology department in 1902. Shortly after, he entered politics as a deputy in the Portuguese Parliament, was appointed professor of neurology in the University of Lisbon in 1911, became Minister of Foreign Affairs in 1918 and then the Portuguese Ambassador to Spain until 1920 when he gave up politics.

As a neurologist and psychiatrist Egas Moniz was acquainted with the use of sodium iodide for treating neurosyphilis. In June 1927 Moniz first performed **arterial encephalography**, by making radiographs of the skull and brain immediately following an injection of sodium iodide solution through a hollow needle inserted into a patient's carotid artery in the neck. He was able to map the major cerebral arteries and their branches within live brains and to define the abnormalities produced by the presence of tumours. In 1928 he presented a cardinal paper on *'Arterial Encephalography, its Importance in the Localization of Brain Tumours'* to the Academy of Medicine in Paris. The method, using different types of contrast media, less irritant than sodium iodide, became a standard neuroradiological investigation throughout the world and generally known as **cerebral arteriography** or **cerebral angiography** because it demonstrated the cerebral veins as well as the cerebral arteries.

In 1936 Egas Moniz devised the neurosurgical technique of therapeutic prefrontal leucotomy for the treatment of certain psychoses. He was shot by one of his psychotic patients in 1939 but survived to receive the Nobel Prize for Physiology or Medicine in 1949, for the invention of leucotomy.

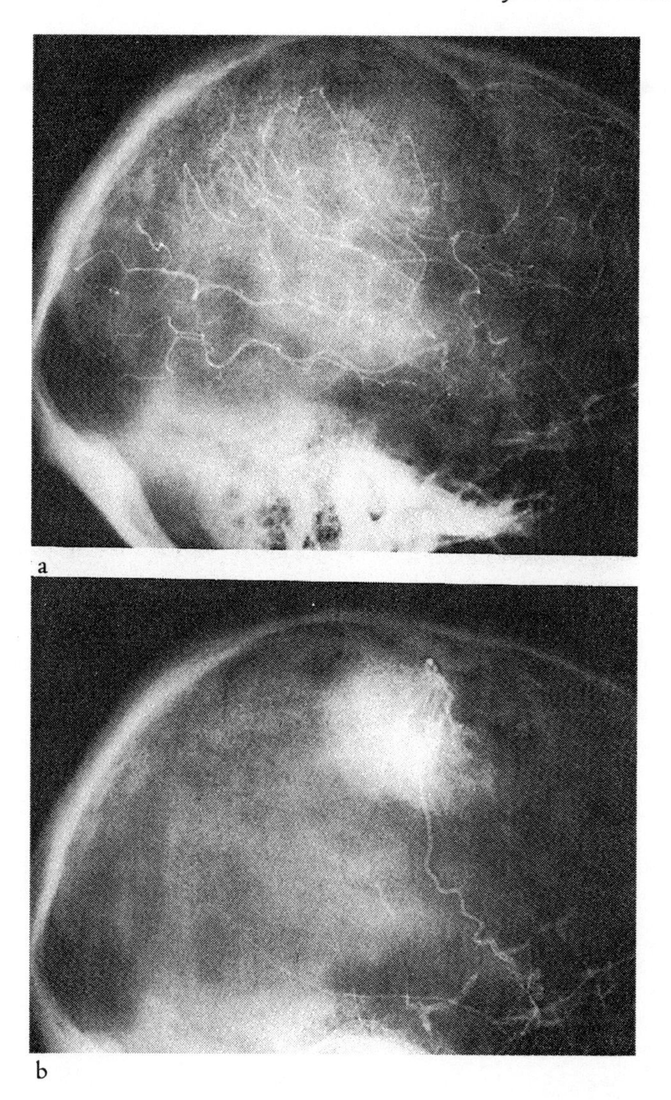

a

b

39. Cerebral angiogram (a) shows distorted cerebral blood vessels due to a **meningioma,** a benign intracranial tumour **External carotid angiogram (b)** shows contrast **blush** within the vascular tumour

In the process of inventing *'arterial encephalography'* Egas Moniz had also devised a technique for opacifying the brain **arteries** and it was not long before others used intra-arterial injections of contrast media to image other arteries in the body. Direct injection of contrast liquid, via a needle stabbed through the skin (*percutaneous*), into the femoral artery opacified the arteries of the leg so as to map them with **percutaneous femoral** or **peripheral arteriography.**

One of the first to expand the use of arteriography was Reynaldo dos Santos (1880 – 1970) a professor of surgery at the University of Lisbon with a special interest in urology. In 1929 dos Santos devised a method for opacifying the abdominal aorta and the renal arteries by injecting contrast medium via a long needle inserted percutaneously from the patient's back into the aorta. This became known as **percutaneous translumbar aortography.**

Dos Santos had experience as a war surgeon and this undoubtedly gave him the confidence to be the first to perform this extravagant procedure. Perhaps he got the idea from watching bullfights? It was surprisingly easy to perform translumbar aortography, under either general or local anaesthesia It continued to be used in patients with blocked femoral arteries due to severe arteriosclerotic disease for nearly fifty years after its invention.

About half a litre of blood leaked from the puncture hole in the aorta after the percuteneous translumbar needle was removed and so patients experienced weakness and back pain after the test; however the mortality rate in such a high-risk group of patients was surprisingly low. Since 1990 **digital subtraction arteriography** that allows computer enhancement of faint images of the aorta, after intravenous injection of contrast medium, has mercifully made direct translumbar aortography obsolete.

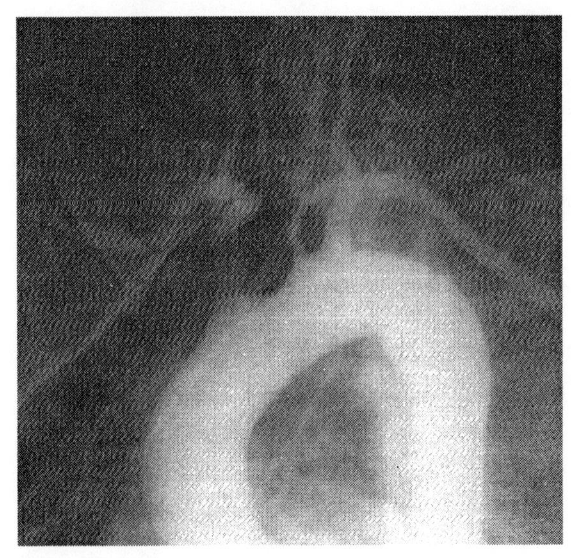

40. Arch Aortogram. A Seldinger catheter has been advanced from the femoral artery in the groin so that its tip is in the ascending thoracic aorta, above the aortic valve. Injected contrast medium outlines the aortic arch, and the carotid and vertebral arteries supplying the brain. The subclavian arteries to the upper limbs are also shown.

Arteriography and aortography following Moniz and dos Santos, relied on direct needle puncture and only a few arteries were easily accessible. Then in 1953 a Swedish radiologist at the Karolinska Institute in Stockholm, Sven-Ivar Seldinger (1921 – 1998), invented the technique of **selective arteriography.**

In this examination a radiopaque polythene catheter is advanced upwards into the abdominal or thoracic aorta, under observation with a fluoroscopic image intensifier. The tip of the catheter is shaped so that it can be manipulated into one of the many arteries that originate from the aorta: **cerebral and vertebral arteries** to the brain; **subclavian arteries** to the upper limbs; **coeliac and mesenteric arteries** to the liver, pancreas, spleen and gut; **renal arteries** to the kidneys and adrenal glands. Selective arteriography greatly improved the pre-operative diagnosis of tumours and arterial diseases involving all these organs.

Before 1956 some centres used *thorium dioxide* suspension instead of iodine contrast media for arteriography, because it had no immediate painful side effects and gave excellent radiographic contrast due to its high atomic number [Z=90]. But unfortunately, as the Curies had discovered in 1898, thorium is radioactive; moreover once injected it is not excreted from the body like iodine compounds, but concentrated in the bone marrow, spleen or liver. Many patients injected with thorium later died of leukaemia, aplastic anaemia or liver cancer and its use was forbidden in 1960. Since 1980 non-ionic, iodine contrast media have been used for arteriography, venography and myelography; they are safe and well tolerated by patients.

Veins are simpler to examine radiographically than arteries because pressure within them is low; they are more accessible and easy to inject with contrast medium. The major reason for examining veins in legs is to demonstrate blood clots within the deep veins (deep vein thrombosis: DVT), which may come loose and pass upwards through the vena cava to the right side of the heart; then into the pulmonary arteries where they may suddenly block the circulation and cause heart failure or sudden death. Over the last ten years, **venography** has been used less often because ultrasound examination of the veins in the legs is quicker and easily repeatable.

ANGIOCARDIOGRAPHY

For centuries **the heart** was regarded as a forbidden territory by surgeons. Then as surgeons and anaesthetists operated on patients with chest diseases such as lung abscesses and TB, the techniques for opening the chest improved. Towards the end of the 19th century a number of patients had survived after surgical suture of cardiac puncture wounds. In the First World War, bullets and shrapnel fragments were removed from the chambers of beating hearts. Heart surgery had become feasible.

In 1929 Werner Forssmann (1904 – 1979) was a cardiology intern who had just qualified in medicine from Berlin University. One day he made an incision into a vein in his arm and fed a catheter up his arm veins and into the right atrium of his heart. Then he walked to the hospital radiology department where he had a chest radiograph showing the catheter tip within his heart. Despite this courageous dedication to medical research, his senior colleagues were outraged; he lost his job and abandoned cardiology for urology.

In 1945, two physicians at Columbia University and Bellevue Hospital, New York City, André F Cournand (1895 – 1988) and Dickinson W Richards (1895 – 1973), adopted Forssmann's technique of **cardiac catheterisation** to pass catheters into the right atrium, then the right ventricle and pulmonary arteries to measure pressures, blood flow and oxygen concentrations in patients with rheumatic and congenital heart disease. Rheumatic heart disease mainly affects the valves of the heart so that it fails to function efficiently as a pump, which causes chronic heart failure. Most common congenital heart defects allow blood to flow abnormally between the two sides of the heart, so that it does not circulate properly through the lungs. As a result blood going to the tissues of the body is poorly oxygenated. In other words a 'blue baby' often has a 'hole in the heart'.

Cardiac catheterisation was a valuable tool for investigating patients with these disorders when cardiac surgery developed techniques for correcting congenital and valve abnormalities in the 1950s. It was an enormous advance for cardiology and heart surgery. In 1956 Forssmann, Cournand and Richards were awarded the Nobel Prize for Physiology or Medicine.

As the techniques of cardiac surgery, anaesthetics and cardiopulmonary support progressed, **angiocardiography** also developed so as to provide surgeons with necessary preoperative assessments. Intravascular catheters positioned under image intensifier control and large volumes of contrast medium were used to examine all the cardiac chambers and adjacent great vessels with dynamic imaging; initially with cine-radiographic cameras, then with videotape and videodisc. Diagnostic angiocardiography was essential for pre-surgical planning of operations for complex congenital heart defects and for assessing failing heart valves.

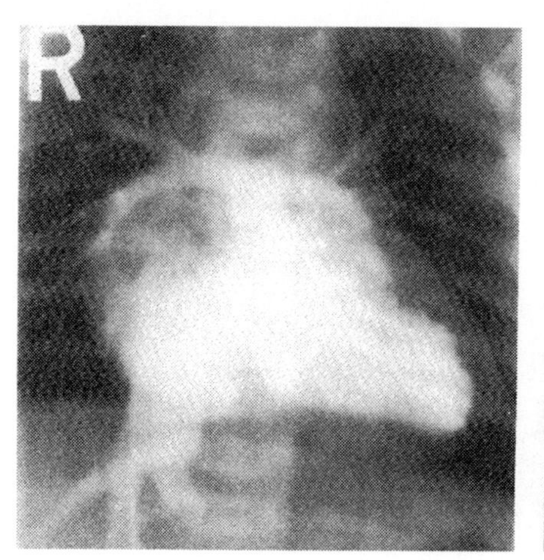

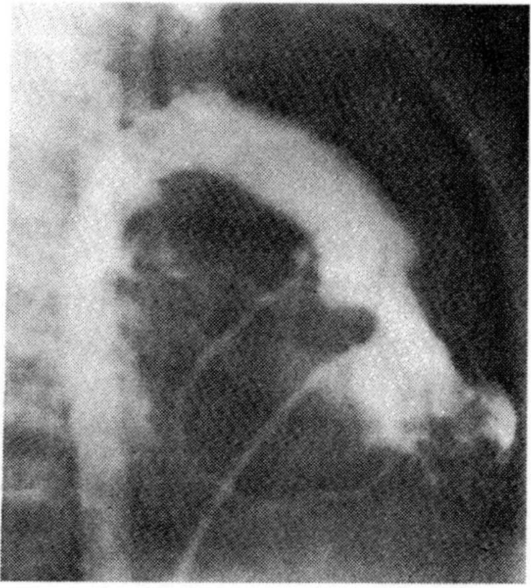

a b

41. Angiocardiography showingTransposition of the Great Vessels. Contrast injected via a catheter in the left ventricle (a) goes into the pulmonary arteries instead of the aorta. Contrast injected in the right ventricle (b) goes straight into the thoracic aorta instead of the pulmonary arteries. Large septal defects between the right and left ventricles allow some oxygenated blood to pass into the aorta, which just keeps the baby alive, but very 'blue'. This congenital heart defect can now be corrected surgically.

The remaining challenge was to visualize the coronary arteries, which supply blood to the heart itself. On 30th October 1958 at the Cleveland Clinic, Dr Mason Sones was examining a 26 year-old patient with rheumatic disease of the aortic valve, which lies between the left ventricle and the aorta. He performed angiocardiography to assess the severity of blood regurgitation through the diseased valve. As contrast medium was injected into the root of the aorta, the tip of the injecting catheter inadvertently entered the origin of the right coronary artery and so the first known **coronary arteriogram** was performed and recorded cinegraphically.

Sones then designed special catheters for entering either the right or left coronary arteries and developed a reliable method of investigating the coronary circulation. In 1962 he published a classic paper entitled *'cine coronary arteriography'* in an American cardiac journal. In 2005 **coronary arteriography** was the most frequently performed angiocardiographic investigation.

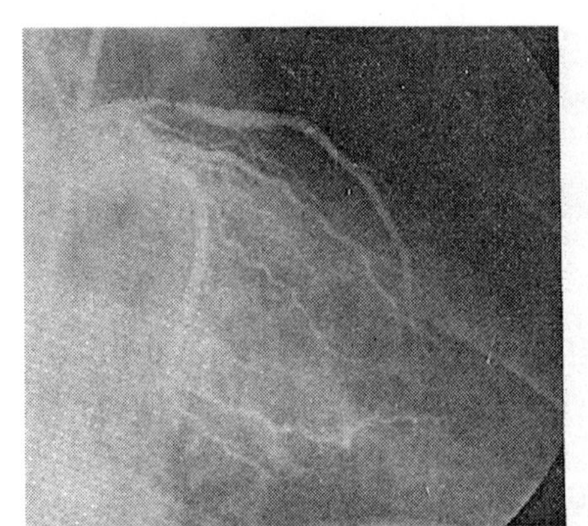
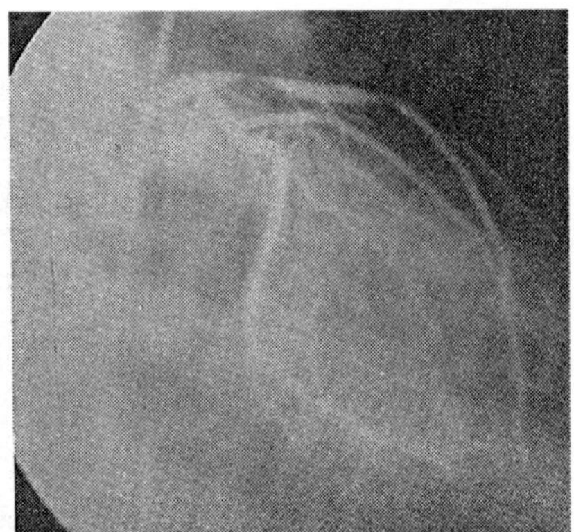

42. Cine- coronary angiogram: two frames show atheromatous narrowing of the anterior descending branch of the left coronary artery.

A few years after Sones' publication, Dr Charles Dotter at the University of Oregon, designed an arterial catheter with an inflatable balloon at its tip, which could be inflated to dilate arterial *stenoses* (pathological narrowings), usually due to atheroma, under angiographic control. This technique, known as **percutaneous angioplasty** is used successfully in arteries such as those feeding the lower extremities and kidneys; without the need for an open operation. It has also proved successful for coronary artery disease and approximately 650,000 coronary angioplasties were performed worldwide during 2005.

Contrast examinations that use iodinated media, such as urography and angiography, have been performed ever since Egar Moniz first injected sodium iodide into the carotid arteries of his patients, eighty years ago. Modern contrast media are much safer but some patients still have reactions to them. Most reactions are mild but life-threatening convulsions, bronchospasm or cardiovascular collapses occur about once in every 2,000 injections, with a death rate of approximately 1 in 100-200,000. Full cardiopulmonary resuscitation must be immediately available whenever these drugs are injected.

Arterial catheterization under local anaesthesia, is unnerving for patients, who usually need sedation. Advancing catheters through atheromatous arteries runs the risk of dislodging atheroma or clots, which may pass further down an artery and completely block it; bad in any artery but life threatening in cerebral or coronary arteries.

Arteriographic procedures are invasive and pose risks. They should be performed by trained physicians with adequate equipment and support and only be used when absolutely necessary.

Despite the risks and shortcomings, urographic and angiographic procedures have produced spectacular improvements in diagnosis so that urologists, neurologists, vascular and cardiac surgeons know the pathological anatomy of their patients when planning their surgical procedures. In many areas radiological interventions, such as angioplasty, can obviate the need for open surgery.

NUCLEAR MEDICINE AND GAMMA IMAGING

Naturally occurring radionuclides (radioisotopes) are those of heavy metals such as lead, radium and thorium. They do not participate in normal physiological processes and so are difficult to use as biological tracers for medical imaging.

In 1934, shortly before the death of Marie Curie, her daughter Irène (1897 – 1956) and son-in-law Frédéric Joliot-Curie (1900 – 1958) bombarded the nuclei of non-radioactive elements with alpha particles and produced the first artificial radionuclides; of nitrogen, phosphorus and aluminium. The Joliot-Curies were awarded the Nobel Prize for Chemistry in 1935 for their discovery.

Very soon afterwards, Ernest Lawrence (1901 – 58), an American physicist who had invented the **cyclotron**, which accelerated protons to very high energies, used his new machine to produce radionuclides of sodium, phosphorus and iodine. These are biologically relevant and can be used as radioactive tracers. In 1939 Lawrence received the Nobel Prize for Physics for the invention and development of the cyclotron and for results obtained with it, especially with regard to radioactive elements. In the same year Frédéric Joliot-Curie demonstrated the uptake of iodine-131 by the thyroid gland.

Before World War II, radioactive materials such as radium or iodine-131 were only used therapeutically: almost exclusively by radiotherapy physicians and medical physicists. Radioactive isotopes or radionuclides were first used diagnostically in medicine in 1942 when George de Hevesy, who had pioneered the use of radioactive tracers in biological research, labelled human red blood cells in the laboratory so as assess blood volume in humans.

The United States government continued its nuclear weapons programme after bombing Hiroshima and Nagasaki; then those of Britain, France and the Soviet Union followed it by building nuclear reactors to make plutonium for their own atomic bombs. The production of artificial radionuclides for biological research or nuclear medicine was a peaceful by-product of the military weapons programmes and so large-scale production of radiopharmaceuticals became feasible.

Hevesy established a principle for the use of **biological and medical radioactive tracers** that - ***"radionuclides and radiopharmaceuticals should participate in biochemical and physiological processes in the same way as the natural chemicals they replace; in other words, a radionuclide should be metabolised by the body just like its stable form."***

Using this principle, new clinical laboratory techniques were devised to assess the function of the thyroid gland by its uptake of iodine-131; body electrolyte composition with sodium-24 and potassium-42; blood cell survival times with chromium-51 and the absorption of iron-59 from the gut. Geiger and scintillation counters gave quantitative results, which could be used to monitor physiological processes within the body over time.

Nuclear medicine investigations became established in chemical pathology laboratories well before gamma ray emitting radio- nuclides could be used for diagnostic imaging.

One of the first attempts to image the thyroid gland in the neck, with iodine –131, which emits gamma rays, was made at the Royal Cancer Hospital, in London, by the professor of medical physics, Val Mayneord in 1950. Geiger counters were moved automatically in a regular rectilinear path in front of the neck of a patient with thyroid cancer after a large therapeutic dose of radioactive iodine had been given. Iodine-131 emits beta as well as gamma rays with photon energies of 364 keV, but only gamma rays can be used for imaging. Although satisfactory images were obtained on this occasion, the technique could not be used for ordinary diagnosis because of the large amount of radiation required.

Better scintillation detectors with sodium iodide crystals were more sensitive and by the 1960s automatic rectilinear scintillation scanners for thyroid imaging were available that required much smaller doses of radioactive iodine than those used for cancer treatment. The technique was then developed for locating tumours in deep-seated organs such as the liver, lung and brain.

In 1958 Hal Anger first displayed the gamma camera, which bears his name, to a meeting of the Society of Nuclear Medicine in Los Angeles. The Anger camera, which in modified form is still used, contains a large sodium iodide crystal backed by an array of photo multipliers.

A heavy lead collimator in front of the crystal focuses photons from within the patient onto the crystal and also protects it from scattered radiation. The shielded camera is placed over the target organ to detect radiation emitted from within that organ after the appropriate radiopharmaceutical has been administered. Gamma photons that strike the crystal are converted into scintillations of visible light; these are detected and enhanced by a photo multiplier array and manipulated electronically to map the distribution of radionuclide in the target organ.

Gamma imaging differs from x-radiography in several ways. The detected radiation is emitted from within the patient's tissues so that gamma images are **emission images** rather than transmission images. Gamma rays are emitted when radioactive nuclei disintegrate and so immediately after the initial intravenous injection of radiopharmaceutical, the whole body receives a small radiation dose before the target organ takes it up. Radioactivity continues as long as radiopharmaceutical persists within the patient.

The radiopharmaceutical delivers a higher dose of ionising radiation to the target organ and to organs in the excretory pathway, usually the kidneys and bladder, than other organs; so the patient and their urine are radioactive for some time after the examination. The rate of physiological excretion of the pharmaceutical and the radioactive half-life of its radionuclide are both important factors in determining the patient's radiation dose.

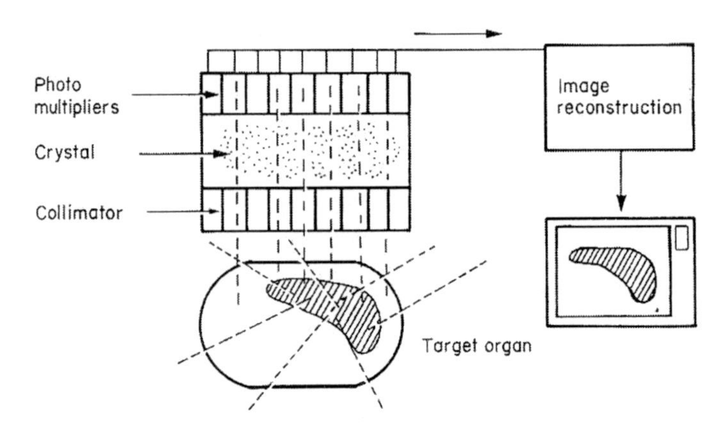

43. Anger Gamma Camera. Photons emitted by radionuclides localized within a target organ are located by a collimated detector crystal; signals are amplified by photomultipliers and mapped by an image reconstruction processor.

Desirable properties for a radionuclide are that:

1. It should have a **short half-life** of just a few hours; sufficient time to allow for chemical attachment to its pharmaceutical vehicle before injection but not too much longer.
2. **Decay to a non-radioactive nuclide.**
3. **Emit gamma rays,** which give images but **no** alpha or beta particles, which do not.
4. Emit **photons with energies between 50 and 300 keV** to which gamma cameras are most sensitive. These properties all tend to minimize the radiation dose to patients.

Technetium-99m fulfils most of the above criteria because it has a radioactive half-life of 6 hours and is a pure gamma emitter with photon energy of 140keV [2]. It combines easily with many useful pharmaceutical vehicles and is used in about 90 percent of gamma imaging investigations because of its flexibility and low radiation dose.

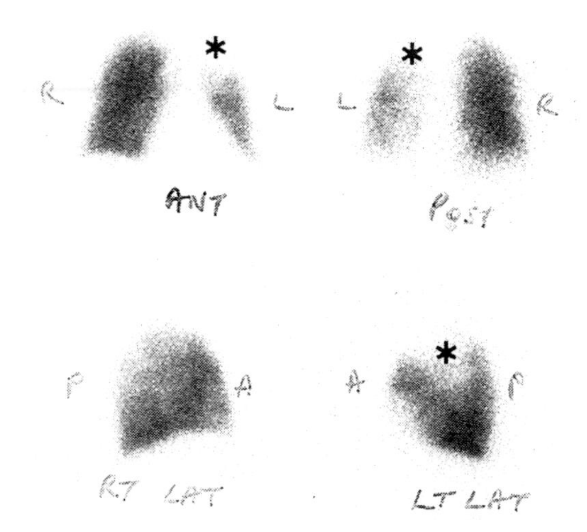

44. Perfusion lung scans with technetium-99m labelled microspheres. There is a perfusion deficit (*) in the left upper lung caused by a pulmonary artery embolism (clot).

The chest radiograph is poor at showing changes of obstructive pulmonary disease whereas gamma imaging of the lungs whilst breathing xenon-133 or krypton-81 gas maps the pattern of ventilation and shows regions of obstructed airways disease, such as emphysema, not evident on the CXR. Pulmonary emboli also produce few changes in the radiographic lung fields but pulmonary blood perfusion scanning with technetium-99m labelled particles (microspheres) reveals regions in the lungs where blood flow has been blocked by emboli (clots) from thrombosed (clotted) deep veins.

Technetium labelled phosphonates can demonstrate primary and secondary bone tumours at an early stage when skeletal radiographs may appear quite normal. Radionuclide bone scanning is especially useful for patients with breast or prostate cancer; to find early metastases and treat them before they cause pain or fractures.

Many diagnostic radiopharmaceuticals are now available for imaging all the major organ systems in the body [3]. Because nuclear imaging procedures reflect physiological processes, they are frequently complementary to radiographic images.

An essential requirement for gamma imaging is the ability to inject the smallest dose of radionuclide that will provide a diagnostic scan. This limitation has an unfortunate consequence for gamma imaging that relates to Schrödinger's cat. It is not possible to know when a radioactive atom will disintegrate so that gamma photon emissions from within a patient are statistical and irregular. With a minimum of radioactivity, relatively few irregular photons are emitted so the gamma camera image is mottled. This is known as *quantum mottle*. For this reason some unkind people refer to nuclear imaging as 'unclear imaging'.

Larger doses of nuclide would give much clearer images; however gamma imaging is not about fine anatomical detail. Its purpose is to map physiological and biochemical processes within a living patient, sufficient for diagnosis but at the same time minimizing the radiation dose.

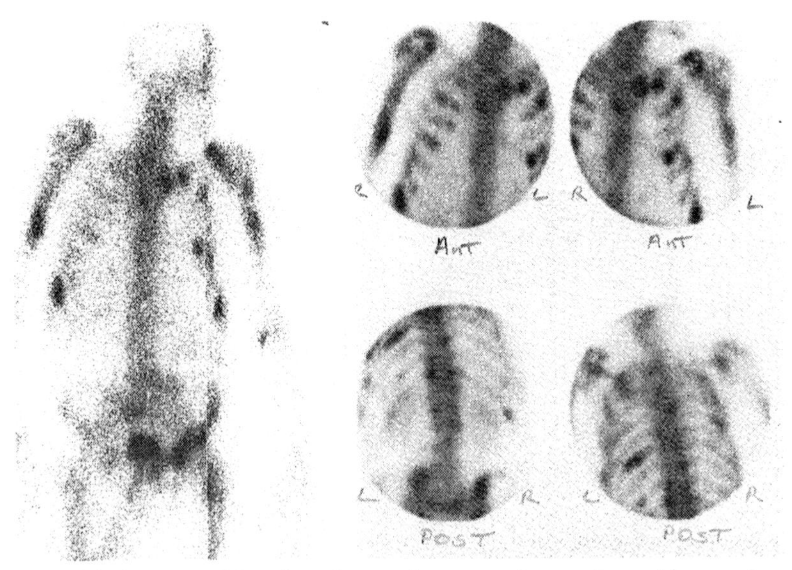

45. Technetium-99m phosphonate bone scan in a patient with secondary **'hot' metastases** in the ribs, arms, spine and pelvis from primary prostate cancer.

THE BIOLOGICAL EFFECTS OF IONIZING RADIATION.

When x-ray and gamma ray photons pass through human tissues, some of them eject electrons from tissue water and other molecules; their photon energy is absorbed by ionization of living tissue. It is **ionization** of tissues that causes radiation injury, which Röntgen warned about within a year or so of his discovery. This was proved when the early radiographers and their patients suffered terrible radiation injuries days, months or years after first exposure.

The absorption of radiation by biological tissues was studied by the physicist and radiobiologist, Harold Gray (1905–65). He was an outstanding nuclear physicist who graduated at the age of only18, then worked at the Cavendish Laboratory with Rutherford, Chadwick and Lawrence Bragg. He investigated the effects of every known type of radiation, including cosmic rays, on different materials, and developed the concept of ***kerma*** (kinetic energy of radiated material) to define the absorption of radiation energy.

After several years at the Cavendish, Gray moved to the Mount Vernon Hospital and Radium Institute, a major cancer treatment centre in north-west London, where he collaborated in landmark research on the effects of radiation on biological tissues and cancers. He found that fast neutrons were ten times more destructive and alpha particles twenty times more destructive than x-rays or gamma rays, for the same *kerma*. These discoveries significantly improved the effectiveness of radiotherapy and other cancer treatments. Gray was a pacifist who refused to take part in military nuclear research during World War II. In 1954 he started the first radiobiology institute in the world at Mount Vernon Radium Institute, now known as the LH Gray Laboratory of the Cancer Research Campaign. He became president of the British Institute of Radiology and his name has been given to the SI unit of absorbed radiation dose, the ***gray* (Gy)** [4].

Human tissues contain 70 – 90 percent water. When irradiated, an ionized water molecule emits an electron and becomes an abnormal positively charged ion: H_2O^+. This immediately releases free radicals: powerful oxidizing agents that produce serious chemical abnormalities in the delicate molecules of tissue proteins and DNA. These biochemical changes in body tissues occur within microseconds of exposure but may take hours, days or even years to become manifest.

The biological effects are shown by changes in the microscopic appearance of chromosomes, which display deletions, abnormal rings and bridges between them. Irradiated tissue suffers impaired growth or may be killed and there is an increased risk of malignancy; gonadal tissues may be sterilized. As Gray had shown, the likelihood of such damage increases with the biologically effective dose and is much greater for fast neutrons and alpha particles.

The biologically effective dose of absorbed radiation energy is measured in ***sieverts (Sv)*** after Rolf Sievert (1896 – 1966), a Swedish radiobiologist who was expert on the effects of low radiation doses and founded the International Society for Radiation Protection [5].

Humans are exposed to many sources of ionizing radiation throughout our lives, from cosmic radiation ; radiation from our environment such as radon from granite rocks; gamma rays from the ground and building materials and radionuclides within our own bodies, such as potassium-40 [6].

Natural background radiation varies with geography and frequent flyers can double the amount of cosmic radiation they receive [7]. Man-made radiations include radioactive fall-out from atmospheric testing of atomic bombs and nuclear accidents, such as Chernobyl, **but the major contribution of man-made radiation to the average citizen in America or Europe is made by diagnostic radiology [8].**

Ever since the 'radiological martyrs' of the early years, radiologists and radiographers have protected themselves with lead aprons and barriers; their day-to-day radiation exposure is monitored with individual dosimeters. Diagnostic radiography for many patients is an occasional experience and most tests give doses comparable with annual natural radiation. Moreover background radiation involves the whole human body whereas radiography usually involves only a restricted region. A major exception to this was obstetric radiography in which the whole baby was irradiated. Moreover a growing foetus is full of rapidly dividing cells; and these had been identified as being more susceptible to radiation, by Claudius Raynaud as far back as 1904.

In 1961 Dr Alice Stewart in Oxford compared rates of childhood leukaemia for those whose mothers had received obstetric radiography during their pregnancy and those who had not. She discovered a 50 percent increase in the incidence of leukaemia in those who had been irradiated *in utero*; that is an increase from the 'natural risk' of about 1 in 3,000 to an 'x-ray risk' of 1 in 2,000. Radiologists and obstetricians did not welcome these results because the findings questioned whether obstetric x-ray procedures were clinically justifiable. However other researchers soon confirmed her findings. In 1968 after a larger study, Dr Stewart showed a 50 percent increase in rare childhood tumours of the brain and kidneys as well as leukaemia. Obstetric radiography was abandoned within a few years and replaced by ultrasound imaging.

It was well known that large amounts of radiation could induce cancer but before 1960 the risks of small doses had been considered as negligible. At about this time it became possible to match known radiation dose levels with risks of malignancy from 15 to 20 year follow-up studies of Hiroshima and Nagasaki survivors. It was accepted that radiotherapy treatments induced cancer in a small number of patients but that the benefits of treatment justified the risk, which was in any case deferred for several years. Dr Stewart's findings were the first to show that even small doses of radiation, such as those used for diagnosis, could be dangerous.

Another diagnostic procedure that was shown to be associated with an increased risk of malignancy was fluoroscopy of female patients with TB, who more often developed breast cancers on the same side as their lung disease; because this side had been fluoroscoped more often. The young female breast, like the thyroid and bone marrow is more susceptible to radiation than other organs.

It has been estimated, by projection from chest fluoroscopy doses, that the risk of inducing fatal cancer is 2 per 100,000 mammography examinations for women less than 50 years and less than 1 per 100,000 for women over 50 years. These projections must be matched against the opportunity of discovering and successfully treating over 200 – 300 breast cancers per 100,000 examinations. However the realization that even low radiation doses might induce cancer further emphasized the importance of protecting patients from unnecessary x-ray tests and keeping radiation doses to a minimum. In the 1960s the need for a non-ionizing imaging method for obstetric diagnosis became especially important.

ECHOGRAPHY: ULTRASOUND IMAGING

As a young man Galileo had performed acoustic experiments with his father, which showed that sound was transmitted as vibrations of air, or pressure waves, initiated by either percussion or vibrating strings. However it was Marin Mersenne (1588 – 1648) a French friar, who was the first natural philosopher to measure the speed of sound with any accuracy. Mersenne did so by timing echoes of his voice exclaiming the seven pulsed syllables of '*Benedicam Dominum!*' from a distant wall, firstly using his pulse and then with a more accurate second-minute pendulum. He was blessed with a result equivalent to 320 metres per second. This is very close to the currently accepted speed of 330 metres per second in air at standard temperature, pressure and humidity. Mersenne remarked that *"sound can fill the sphere of its activity only in a space of time."*

He established the basic principle of echography, that timing pulsed echoes can measure distances; the more precise the timing, the more accurate the measurement.

Echo-sounding was first used in 1925 as a method of measuring the depth of water beneath a ship by using acoustic oscillators, which could emit and detect sound waves in water with a frequency of 20Hz. This frequency is above the human audible range and travels further through water than audible frequencies. A stylus on a rotating paper drum recorded the time from emission to return of echo. As the ship passed over the seabed its contour was mapped. Shoals of fish could also be detected by fishing boats with echo sounders. The naval threat of submarine warfare led to the development of sound navigation and ranging systems (SONAR) to locate enemy submersibles at significant distances from a threatened warship.

Ian Donald (1910 – 1987) qualified in medicine at St Thomas' Hospital in London in 1937 and served with distinction as a medical officer in the RAF during the Second World War. Whilst in the air force he became involved with RADAR (radio detection and ranging) and SONAR.

After the war, he became a university reader in obstetrics and gynaecology at St Thomas' and in 1954 attended a lecture given in London by John Julian Wild, a surgeon from Minnesota, on echo imaging of pathological specimens, especially bowel tumours.

Shortly afterwards Ian Donald became the professor in the University Department of Midwifery in Glasgow and learned that ultrasound probes were used in the city's dockyards to detect flaws in ships' steel boilers. He persuaded the engineers to let him examine surgical specimens of uterine fibroids and ovarian cysts with their instrument. He found that ultrasound passed right through the cysts and echoed easily from their back walls but that it penetrated poorly through solid tumours. Encouraged by the results Donald, with the help of an engineer; Tom Brown, adapted an ultrasound flaw detector for use in his clinic to detect cysts and fibroids in patients, before their operations. He published a paper entitled *Investigation of abdominal masses by pulsed ultrasound* in the *Lancet* in June 1958.

Although ultrasound implies sound with a frequency above 20 kHz, the frequencies used for medical imaging are almost a thousand times higher; between 1 and 20 MHz, and these are created and detected by piezoelectric crystals. In 1894 Pierre Curie and his brother had discovered piezoelectricity, a phenomenon by which an electric field applied across a quartz crystal makes it either contract or expand. Conversely, compression or tension applied to the crystal produces an electric field. A rapidly alternating electric field causes a piezocrystal to vibrate at high frequency and create ultrasound waves. A returning echo from a single short ultrasound pulse causes the crystal to vibrate and induces an alternating field across the crystal. The time interval between emission of a pulse and detection of its echo is determined by the distance of the reflecting surface from the face of the crystal. The strength of the returning echo determines the size or amplitude of the electric signal. A medical ultrasound probe contains a ceramic piezocrystal, which produces and detects sonic pulses; it is also known as a transducer, because it can convert electrical signals into sound waves and sonic pulses back into electrical signals.

The velocity of sound in body tissues is similar to that in water and is approximately 1,500 metres per second. However internal body dimensions relevant for imaging are measured in millimetres rather than metres, so velocity of medical ultrasound is better expressed as 1.5 millimetres per microsecond. The wavelength of 3 MHz ultrasound, which is a commonly used frequency in medical diagnosis, is 0.5 mm.

Modern electronic methods make it possible to time the passage of an ultrasound pulse to within a fraction of a microsecond and so the distance between the transducer and reflective surface can be determined to within a fraction of a millimetre.

Ultrasonic pulses are partially reflected at interfaces between tissues of different acoustic impedance or elasticity [9]. The acoustic impedances for water, blood, or muscle are similar so that ultrasound can pass through interfaces between these types of tissue easily and is partially reflected and partially transmitted at the tissue boundaries between them.

However the impedance of air or other gases is negligible so that their interfaces with body tissues reflect ultrasound completely; tissues in the shadow of air or bowel gas are therefore invisible. For the same reason, good contact between the transducer and the patient's skin using oil or gel to remove any air between them is essential for ultrasound imaging. On the other hand, bone has a very high impedance of and also reflects ultrasound completely. Because cysts contain watery fluid within well-demarcated walls, they are very well shown. Most soft tissue masses can be well defined if they are not too deep within the body but air in the lungs, bowel gas and bones are reflective barriers to ultrasound.

A (amplitude) mode echography. A-mode recording is like a ship's depth sounder and shows the distances of reflecting interfaces from the transducer, which are portrayed on the horizontal axis of a cathode-ray oscilloscope (CRO). Strong echoes give high **amplitude** deflections and weak echoes small deflections. A-mode presents information in one dimension, which is the axis of the ultrasound beam; maximal echoes are received from tissue interfaces at right angles to the path of the beam.

B (brightness) mode echography. If the strength of an echo is displayed as a variation in **brightness** on the CRO instead of a deflection, it is possible to produce two-dimensional B-mode compound images of a section of tissues within the patient by traversing the transducer and the beam through that section.

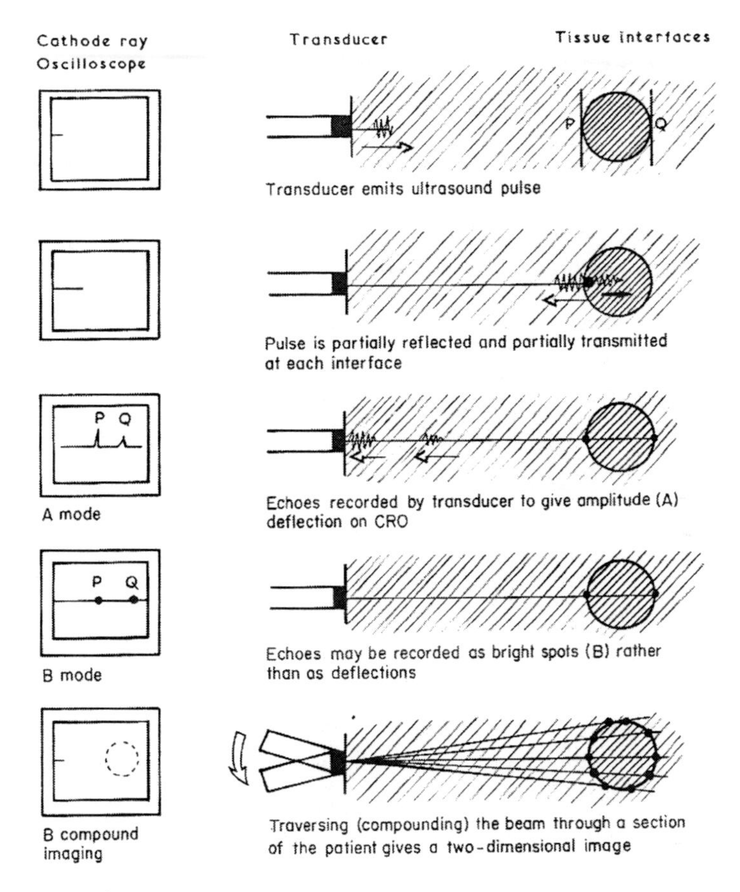

46. Diagram of A mode and B compound mode imaging. The method of reconstructing the compound image on the screen is similar to that for radar screens.

Highly reflective boundaries show as bright structures on the scan and echo-free structures such as fluid in cysts and blood vessels show as dark areas; however the boundaries of cysts and vessels are well demarcated.

Real-Time ultrasound imaging. Modern ultrasound probes are composed of arrays of very small transducers, each rapidly transmitting and receiving pulsed signals, which are computer-processed to create high definition images of internal body structures and their movements as they happen, that is in real-time. Real-time ultrasound can be regarded as three or even four-dimensional. The instruments are easy to use for obstetric, abdominal, paediatric, cardiovascular and ophthalmic applications.

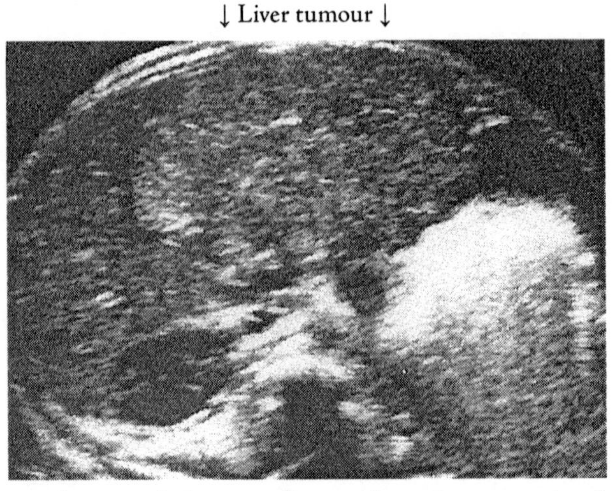

47. **Liver ultrasound** showing a large echogenic tumour involving the liver.

Doppler ultrasound. In 1842 Christian Doppler, an Austrian physicist first observed the change in frequency of sound of a vehicle coming towards and then away from the listener; from a high pitch to a lower pitch. Doppler ultrasound methods depend on a parallel principle that structures moving towards or away from the transducer either increase or decrease the frequency of the returning echo. Red blood cells reflect high frequency sound and so Doppler ultrasound methods can be used to assess velocity of blood flow and in combination with anatomical images, can assess volumes of flow.

Modern real-time scanning, which combines colour-coded (red or blue) Doppler flow imaging with dynamic anatomical imaging (shades of grey), has proved very effective in demonstrating occlusive disease in arteries and veins. Ultrasound vascular imaging has significantly reduced the need for invasive arteriographic and angiocardiographic procedures over the last twenty years.

In 1959 Ian Donald used his probe to examine his pregnant patients and found that he could get clear echoes from both sides of a baby's head. He was able to locate the position of the head and spine within the uterus and the number of heads in multiple pregnancies. It was also possible to locate the position of the placenta. The other marvellous thing about ultrasound was that it was non-ionizing and harmless in small diagnostic doses. **Obstetric echography** had been born; however Donald always called it *sonar.*

Other obstetricians followed his lead and began to use it, whether it was called *sonar, echography or ultrasound imaging.* It was soon found that the width of the foetal head (biparietal diameter) correlated closely with foetal maturity, a factor that is important for good obstetric care. It was also helpful in early pregnancy for assessing the presence of a live embryo by its cardiac motion, evident as early as 6-7 weeks. Gantry mounted probes were developed to give 2-dimensional, B-compound imaging of the pregnant uterus and then of other abdominal organs. The emergence of obstetric ultrasound imaging could not have come at a better time as obstetricians began to accept Alice Stewart's evidence of the hazards to babies of obstetric x-rays.

Ultrasound offered a safe alternative. By 1966 a number of manufacturers in Europe and America were making machines to satisfy the new ultrasonic boom.

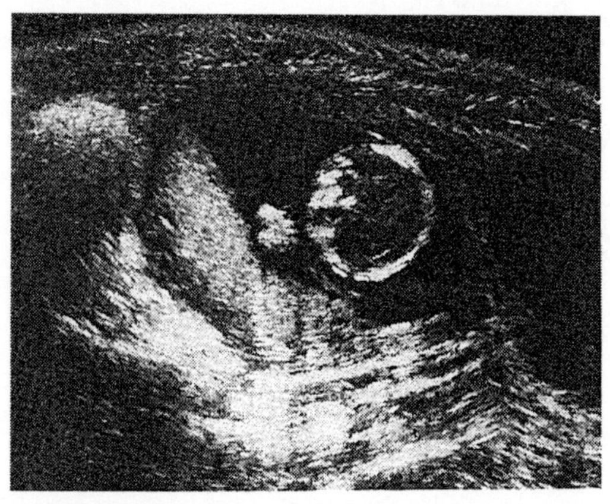

48. Obstetric ultrasound at 18 weeks maturity showing the diameter of the foetal cranium and the position of the placenta

Technical improvements in ultrasound have made detailed foetal assessments routine and are performed at about 18 to 20 weeks maturity, so as to assess foetal health and to identify serious congenital abnormalities. The implication that the mother must be offered an abortion if serious deformities are discovered is an ethical issue posed by ultrasound. As a physician who had taken the Hippocratic Oath, Ian Donald was opposed to abortion for apparently normal embryos and opposed British abortion legislation in 1967.

Ultrasound imaging has transformed clinical diagnosis over the last forty years, not only in obstetrics and gynaecology, but also in abdominal, cardiac and vascular imaging and endocrinology. Small internal probes have been designed for trans-vaginal examination of the uterus and ovaries, trans-rectal examination of the prostate and trans-oesophageal echocardiography.

Ultrasound is intrinsically non-invasive and does not use harmful radiation for diagnostic use. Worries that it might be shown, like x-rays, to have long term effects on children examined *in utero*, have been dispelled by follow-up studies of millions of examinations. Because ultrasound imaging maps entirely different physical properties of body tissues compared with ionizing radiations, it is so often complementary to radiography or nuclear imaging. Ultrasound has replaced invasive x-ray procedures for solving many serious diagnostic problems, and has made substantial contributions to investigative medicine. Its major limitation is that air, bowel gas and bones are barriers to sound waves.

COMPUTER-ASSISTED IMAGING

Signals from the TV camera of a fluoroscopic image intensifier are analogue signals: voltages, which vary smoothly as the image on its output phospher is scanned into a television raster of horizontal lines. If analogue signals are converted to digital signals with an A-D converter, images can be processed in a variety of ways. Modern silver less x-ray units also produce digitised images by laser scanning a radiographic phosphor, which can be processed by a computer to yield more diagnostic information. This computer-assisted technique is known as **digital radiography**.

Arteries with dilute amounts of iodine in their blood after intravenous injection of contrast are very faintly radiopaque. These arteries can have their faint images enhanced and overlying shadows removed so as to make them diagnostically useful with **digital subtraction arteriography**. A digital image A is made before the intravenous iodine contrast injection, and then digital image B is taken in exactly the same position after contrast injection. Because the image data is digital, a computer can subtract A from B to make a subtraction image of the radiopaque arteries. Arteries can be imaged without having to be punctured. Unfortunately if the patient moves between A and B, digital subtraction doesn't work.

A new type of digital subtraction scanner, called a **synchrotron,** can detect the unique quantum electrodynamics of iodine atoms to overcome motion problems [10]. A synchrotron scans the body with two x-ray beams simultaneously:one beam with higher energy than the other. The lower energy image is subtracted from the higher energy image to make a synchronous subtraction image of arterial iodine that is not blurred by movement [10]. With a synchrotron it is possible, after a small intravenous injection of contrast, to image moving coronary arteries without arterial catheterization.

COMPUTER-ASSISTED TOMOGRAPHY

Conventional radiographic images superimpose *3-dimensional structures* into *2-dimensional projections* or *shadow pictures,* with the risk that overlying anatomy obscures pathology. Attempts were made to overcome this problem by swinging an x-ray tube in such a way that the primary beam always centred on the same level inside the patient. This imaging method focused on one anatomical section and blurred out the detail of structures in front and behind this section.

Repeated swings focused at different depths had the effect of slicing the body into successive thin sections for localized imaging.

The resulting blurred images gave approximately 2-dimensional images of notional 2-dimensional slices of tissue. The method is still used occasionally and is called **conventional tomography** [*f* Gk. *tomos:* slice or section + -graphy].

Conventional tomography was most successful for the lung to discriminate between benign and malignant pulmonary masses. It did not give clear images and the whole chest had to be irradiated to image each slice. Making multiple slices through the chest involved much more radiation than a single CXR. Images were often unsatisfactory, especially when tomography was used for bones rather than lungs.

Computer signal processing became a force in medical imaging after 1972 when it was first used with x-rays, then with gamma rays, to produce computer-assisted tomograms of the human body. Similar methods were later applied to make body images by mapping signals of resonating radio waves in strong magnetic fields (MRI) and from positron emitting radionuclides (PET). Each of these radiations gives out its own type of physical signal; however the different signals are processed using similar computer techniques to map the body in sequential sections and make interpretable images for each of these different imaging modalities.

Computer-assisted [*x-ray*] **Tomography (CT)**
Single [*gamma*] **Photon Emission CT (SPECT)**
Positron [*coincident gamma*] **Emission Tomography (PET)**
Magnetic [*radio frequency*] **Resonance Imaging (MRI)**

For the purpose of tomographic (body section) imaging, the body can be sliced notionally into sections in planes at different angles. The three major orthogonal planes used in medical imaging correlate with traditional anatomical planes; names redolent of mediaeval executions, battles and royal glory.

Axial sections are transverse and separate above from below.
Sagittal sections (*sagitta: l. arrow*) are in planes parallel to the truly **sagittal midline plane,** which separates side from side.
Coronal sections (*coronalis: l. crown*) are at right angles to the other planes; coronal sections separate front from back.

Planar sections or slices of the body to be imaged are notionally divided into many cuboidal blocks of tissue, or volume elements known as a **voxels.** This allows the computer to map digital signals from each **voxel** to a matching picture element (**pixel**) in an anatomical picture.

Voxels are arranged in rows and columns so as to convert each section into a network or **matrix:** typically these rectangular matrices are made up of either **256 x 256** (65,536) or **512 x 512** (262,144) voxels with matching images having the same number of pixels, each mapped to a voxel.

Each section or slice and each notional voxel within it may be 1 – 10 mm in height; and each voxel is usually 1 – 2 mm in width and depth. Varying with modality, a voxel of tissue will either transmit or emit a signal. The signal from each voxel is digitised, computer-processed and then portrayed as a matching picture element **(pixel)** in a 2-dimensional image matrix so as to form an anatomical picture or map of the real 3-dimensional section. The strength of signal from each voxel is expressed as the 'brightness' of each matching pixel to give a **grey-scale image.** This map or picture may be displayed on a television screen, usually as a grey-scale image; however it can be colour-coded, which is useful for gamma imaging.

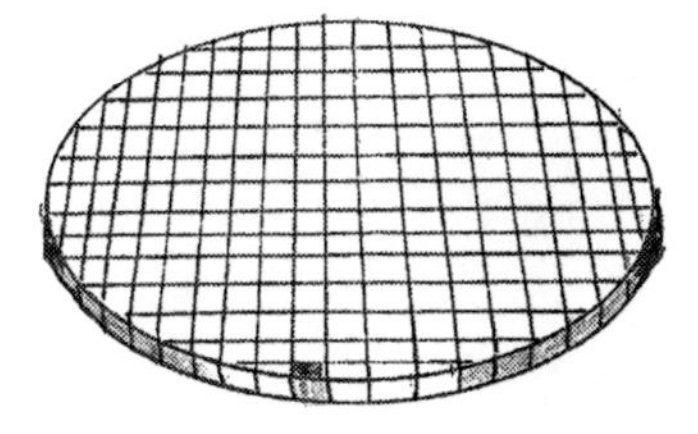

3-D body section of voxels

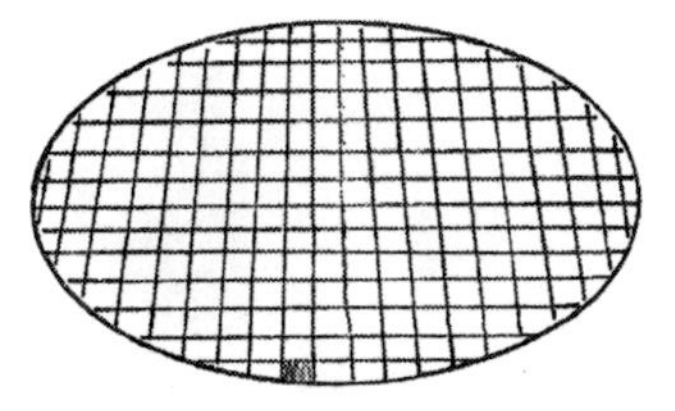

2-D image of matching pixels

49. Voxels and pixels.

The annual meeting of the British Institute of Radiology was held in London in 1972 and included two papers on *Computerized Axial Tomography (CAT) of the Brain,* which were given by James Ambrose, a neuroradiologist and by a shy research engineer, Godfrey Hounsfield (1919 – 2004), who had worked on radar whilst in the RAF during the War and became an electronic engineer afterwards. He joined the EMI company in 1951 to work on radar and rocket guidance systems, and then led a team building one of the first all-transistor computers.

He began to study computerized pattern recognition in 1967 when the concept of CAT occurred to him.

Hounsfield had spent his working life in top-secret research establishments and had little experience of giving public lectures, certainly not to a large audience of radiologists, radiographers, neurologists, physicists and engineers. Before the talks, few in the audience knew what computer-assisted tomography was, because its invention and early research had been kept so secret. As the presentations were made the audience was amazed as James Ambrose showed images of the brain and of brain tumours, and Godfrey Hounsfield explained how the model 1001 CAT scanner worked.

The papers were followed by prolonged applause and soon caused enormous excitement in the medical, physics and engineering communities as the news spread. The images were coarsely pixelated because an **80 x 80** matrix had been used, but this did not matter. For the first time it was possible to image the brain itself: grey matter and white matter, without sticking contrast media into arteries or ventricles. The entire brain was visualized, not just its arteries or the spaces around it.

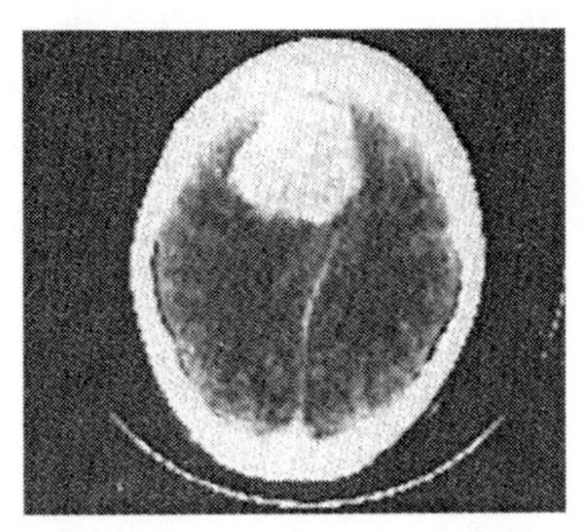

50. CT Brain scan of a Meningioma : the same tumour as shown on previous cerebral and carotid arteriograms

The concept was brilliant because it combined the photon output from a conventional **x-ray tube** with the sensitivity of gamma ray **scintillation detectors**; a **computer** to process digitised signals and a **digital imaging system,** to transform quantitative data into a detailed grey-scale **TV picture**.

An early CT head scanner consisted of an x-ray tube mounted on a gantry with detectors on the other side. The tube produced a beam of x-ray photons with a constant dose rate. As the gantry rotated around the patient in the axial plane, the detectors measured the attenuation of the x-ray beam after it had passed through the head from different angles. The thickness of the axial section was 8 or 10 mm, the same the width as the beam, and 12 sections were needed to include the entire brain [11].

Thousands of measurements were made as the gantry rotated around the patient's head. From these measurements the computer calculated the x-ray attenuation of each and every voxel in an axial section of the head. Hundreds of thousands of equations had to be solved by the computer to plot the **attenuation coefficient** of every voxel, using a mathematical technique known as *filtered back projection.* The attenuation coefficient of a voxel represented the electron densities of the atoms within it [12].

Transposing the results produced a quantitative map of each of twelve sections through the cranium and brain, which reflecting the electron density of every voxel of bone and brain tissue. These were displayed as grey-scale pixels on the TV picture matrix. After a few more years larger CT scanners were designed to examine the entire human body as well as the head.

Attenuation coefficients became known as **Hounsfield units** (HU) and in theory, represent a very wide grey scale with a range of 2,000 shades of grey instead of just the four plain radiographic values of bone, water, fat and gas. But the human eye can only distinguish 16 shades of grey so it is not possible for an observer to assess all the information in a CT scan in one view on a TV monitor. Instead it is necessary to explore the computer's memory by 'windowing' up and down at different HU levels. Lowest to look at air in the lungs, then up to look at the brain, heart, abdominal organs and other fatty or water-density soft tissues: and then further up to look at bone, skeletal anatomy and iodine contrast enhanced blood vessels and cardiac chambers.

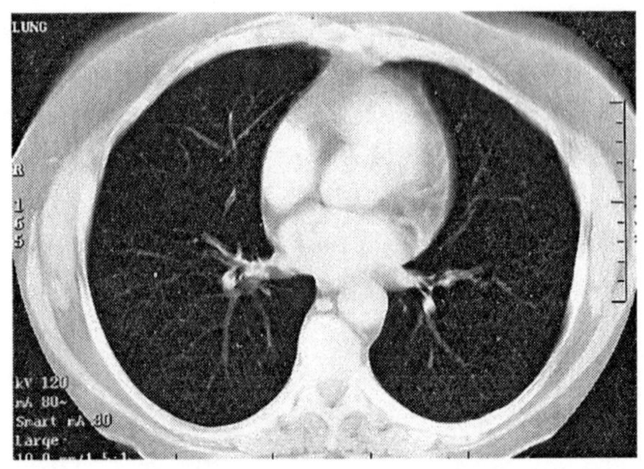

51. Axial CT scan of the chest to show the lungs (window set at – 600 HUs) Air/water contrast is best at this low setting. The pulmonary blood vessels of water density are outlined by air in the surrounding lung tissues. The patient's left side is on the viewer's right.

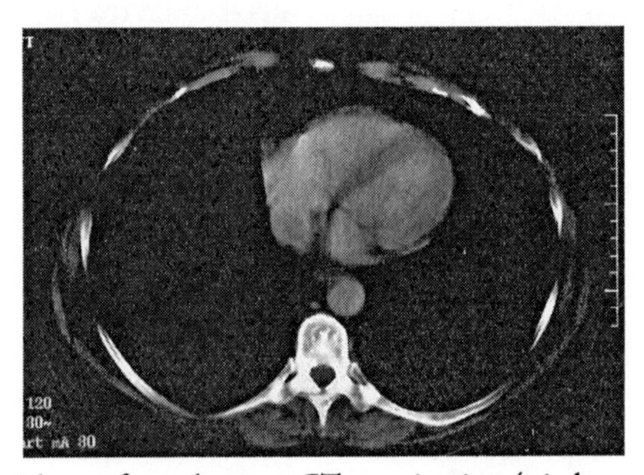

52. An axial scan from the same CT examination (window set at + 80 HUs) shows the cardiac chambers, ribs and spine. Contrast between bone/soft tissue and iodine/soft tissue is best at this higher setting. The cardiac chambers have been opacified with intravenous contrast medium. At this window the ribs and vertebra are more visible than on the lung window. This section is 3 cm below the section shown in 51.

From the beginning, **CT of the brain** was a great success and is now essential for neurology, neurosurgery, accident and emergency medicine. It is easy to perform and involves less radiation to the patient than cerebral angiography or air encephalography; its introduction led to a dramatic reduction in the use of these unpleasant procedures.

CT shows fresh haemorrhages in the brain as well as brain tumours and can distinguish between strokes caused by bleeding (haemorrhage) from ruptured cerebral arteries and those due to clotted arteries. Clots in blocked arteries can be dissolved by prompt thrombolytic treatment, but this cannot be used for cerebral haemorrhages because it will make bleeding from leaking arteries worse. Urgent CT is needed for stroke patients, to make the correct diagnosis quickly and to give the right treatment as soon as possible. Brain CT is also necessary to assess brain contusion and bleeding after serious head injuries.

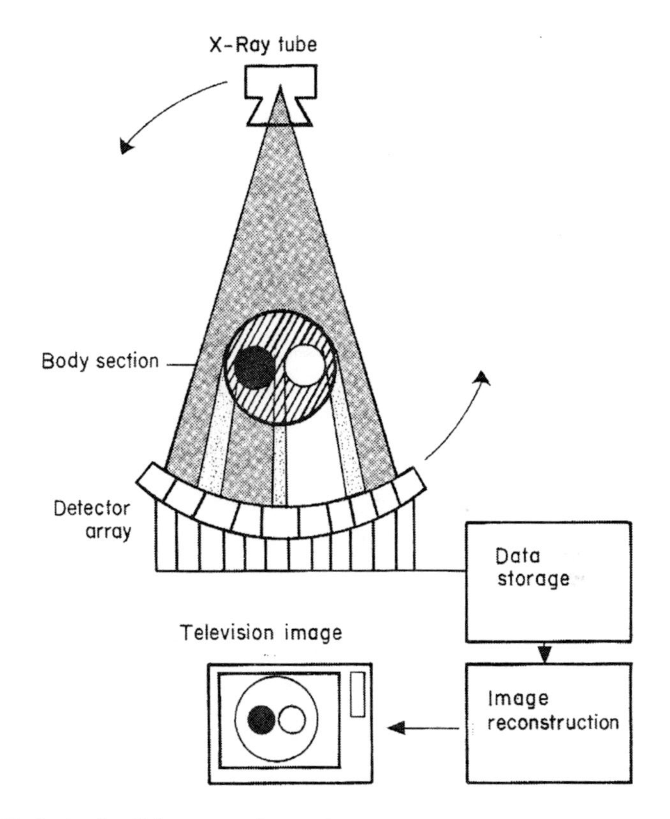

53. Whole-body CT scanner (c1990). The x-ray tube and detector array are mounted on a gantry that rotates around the patient.

Whole-body CT, first became available in 1976 and initially had more problems than cranial CT because body motion caused blurring of images. It was very difficult for patients to hold their breath repeatedly for 20 to 30 seconds during each scan rotation and many more sections were needed to cover the chest or abdomen than the cranium. Modern helical scanners with multiple axial x-ray beams can now scan a patient's whole chest or abdomen in one 10-second breath-hold. A stack of consecutive '2-dimensional' axial sections makes a 3-dimensional block of voxels, whose Hounsfield numbers can be used to reconstruct computer-generated pictures in the sagittal and coronal planes, as well as 3-dimensional images.

Whole body CT is an enormous advance on conventional radiography because it solves the problem of overlapping structures. Tumours and abscesses anywhere in the body can be diagnosed and assessed accurately before surgery. The response of cancers to chemotherapy or radiotherapy can be monitored. A variety of other diseases such as scarring of the lungs, narrowed arteries or aneurysms and many other pathologies, can be diagnosed with ease.

The major limitations of CT are the relatively large doses of ionizing radiation required and the expense. Cranial CT gives about 2 mSv to the head, which is acceptable but whole body CT gives abdominal and pelvic doses, equivalent to barium studies, or about 8 mSv. Significant doses to the radiosensitive ovaries and bone marrow are unavoidable. Pregnancy is a strong contraindication to abdominal CT because of the radiation risk to the unborn child.

CT scanners are expensive. The British National Health Service has generally been unable to provide enough CT scanners because of their cost, except to a handful of major city hospitals throughout the United Kingdom. The large NHS general hospital, which provides most of the medical, surgical and accident services to London's Heathrow airport; the World's largest international airport and only 4 kilometres from Godfrey Hounsfield's laboratory and the EMI factory that made the first CT scanners in the World, did not have a CT machine of its own until 1990. This machine was paid for by a local charity because the NHS could still not afford to 18 years after Hounsfield's brilliant invention.

Godfrey Hounsfield shared the Nobel Prize for Medicine with Allan Cormack, a South African who had independently developed a similar device. He was knighted in 1981.

Computed tomography is a technology that is still evolving. It gives good spatial/anatomical definition but can also give very accurate molecular information. For some years CT has been used to scan airline passengers' baggage to detect terrorist explosives, like *Semtex,* which have unique molecular structures, electron densities and signature Hounsfield numbers. More recently, Howard Chang from Stanford University in California has been studying the DNA profiles of different body tissues in patients with cancer to identify neoplastic molecules using standard CT scanners. It may soon be possible to avoid the need for invasive microscopic biopsies to confirm the diagnosis of cancer.

GAMMA EMISSION TOMOGRAPHY

The axial computerized *filtered back projection* CT technique was soon applied to radionuclide imaging, which shared with radiography the problem of 3-dimensional signals onto a 2-imensional projection image. It was called **single photon emission tomography (SPECT)**; 'single photon' to distinguish it from the 'dual photon' imaging of PET. The benefits were not as great as for x-ray CT because gamma cameras already utilized sensitive scintillation counters and because gamma emissions from radiopharmaceuticals within a patient occurred in the same scattered, probabilistic, Schrödinger way as for an Anger gamma camera.

A SPECT gantry is much simpler than a CT gantry because it only needs to position the detectors as they rotate around the patient in the axial plane; it rotates much more slowly than CT because the photon count rate is much lower. However SPECT can locate some abnormalities better than a conventional Anger gamma camera. For example, radionuclide bone scans use technetium-99m phosphonates that are excreted in urine, so activity in the bladder obscures the pelvic bones on a conventional gamma scan. SPECT overcomes this problem by imaging the skeleton in the axial plane, separating the pelvic skeleton from the 'hot' bladder.

Another important use for SPECT is in **myocardial perfusion imaging** with either thallium-201 or technetium-99m myocardial bonding agents, which are taken up by the heart muscle. In patients with angina, SPECT myocardial scanning at rest and after exercise can identify those parts of the heart that are starved of blood during exertion, because of coronary artery disease. SPECT locates blood deprived areas of muscle better than a gamma camera.

Paul Dirac, who became the Lucasian professor of mathematics in Cambridge, predicted the existence of *antimatter*. His relativistic wave equation indicated that for every species of subatomic particle there had to be another species of the same mass but opposite charge; an *antiparticle,* and he predicted the existence of the *antielectron* in 1928. Dirac's hypothesis was confirmed in 1932 when the **positron** was discovered experimentally.

A positron has a mass equivalent to that of an electron but a positive charge. A positron cannot survive for very long because it soon meets an electron. Both particles are annihilated, releasing energy as gamma photons [13]. The two photons are emitted at 180° to one another known as *dual emissions* or *coincident photons* because they are emitted at the same instant from the same point of origin.

Positron emitting radionuclides, which are produced in a cyclotron, are the basis of **positron emission tomography** (PET), and usually have short half-lives measured only in minutes. **Fluorine-18 labelled fluoro-deoxy-glucose** (**^{18}FDG**), an analogue of glucose, can be used to monitor glucose utilization in the body. It can identify regions of active metabolism, either in the brain or cancers, which burn up glucose at a rapid rate. Fluorine-18 has a half–life of 100 minutes, which allows time to label deoxy-glucose and send the radiopharmaceutical on short journeys to other clinics that do not have a cyclotron. Carbon-11 has a half-life of only 20 minutes so that carbon-11 labelled pharmaceuticals can only be used in clinics that have a cyclotron, a radiopharmacy and a PET scanner.

A PET scanner has a ring of detectors, which identify coincident gamma photons; solitary photons are not registered, and a computer, which plots the path lines of dual emissions. A straight line between two coincidentally activated detectors crosses the point of origin of the positron emission. Repeated dual emissions gradually build up a map of radiopharmaceutical distribution, such as ^{18}FDG, within the patient [13]. The resulting image can be viewed as either a projection or as a section in any orthogonal plane so as to identify metabolic 'hot spots'.

PET images have a low signal-to-noise ratio and so appear very mottled. Interpretation of these physiological images is aided by comparison with equivalent anatomical CT scans. Indeed there are now combined PET/CT instruments, which perform both investigations at the same time, to provide anatomically registered maps of physiological activity. These combined PET/CT machines are extraordinarily expensive but are valuable in assessing the presence of active cancer in tiny nodules detected by CT. Such nodules may be too small for guided-needle biopsy, which is used to prove microscopically that cancer is present, before giving chemotherapy or radiotherapy. Instead, suspicious nodules may be 'biopsied' with positrons, so using PET as an alternative method of proof.

PET scanners can also be used for **functional brain imaging,** to research and monitor brain activity by plotting glucose utilization with ^{18}FDG in different regions, and by mapping uptake of carbon-11 labelled neuro-pharmaceuticals by synaptic neurotransmitter receptors in the cortex, basal ganglia and brain stem centres. Neuroscientists use ^{11}C pharmaceuticals to identify the location and profusion of brain receptors for major neurotransmitters such as dopamine or serotonin in patients with psychoses or drug addiction. Positron emission tomography has enormous potential to monitor metabolic processes throughout the body as well as in the nervous system but is likely to be used for only a limited number of clinical indications or research investigations because of its expense.

MAGNETIC RESONANCE IMAGING

At first it seems impossible that radio waves with wavelengths of metres can be used for imaging anatomical structures to a spatial resolution of a millimetre or less. However chemists have used radio waves and the phenomenon of **nuclear magnetic resonance (NMR)** to study the structure of complex molecules for half a century. Two American chemists, EM Purcell at Harvard University and Felix Bloch at Stanford University developed analytical hydrogen (proton) NMR, for which they received a joint Nobel Prize in 1952. Twenty years later Paul Lauterbur in New York used NMR to make 2-dimensional images of an object by using magnetic field gradients to modify the strong uniform field surrounding it. Then Peter Mansfield in Nottingham used mathematical techniques to process hydrogen radio signals, which made NMR a practical imaging procedure. The Nobel Prize for Physiology or Medicine was awarded jointly to Paul Lauterbur and Peter Mansfield in 2003.

Medical **magnetic resonance imaging (MRI)** was first introduced as a clinical technique in the 1980s. '*Nuclear*' was dropped from its title to indicate to patients that MRI did not involve radioactivity. The body is transparent to radio waves and its tissues can *absorb and re-emit radio signals (resonate)* at specific frequencies, within a strong magnetic field. MRI gives tomographic images which are comparable with x-ray CT.

An **MRI scanner** comprises:

a powerful **magnet** to enclose the body in a strong field;

gradient coils to modify the main field and so *characterize and locate tissue voxels magnetically*;

radio coils to transmit pulses and to receive *resonant radio signals* from specific *voxels* within the body; and a

computer to coordinate magnetic gradients with radio signals and construct anatomical images.

Magnetic fields are induced by flowing electrons in wires or in TV tubes, and by paired electrons that oscillate within the crystal lattices of ferromagnetic metals. Protons also have magnetic properties. Spinning protons create their own magnetic moments and the atomic nucleus with the largest magnet moment is that of hydrogen: a single proton. The human body contains large amounts of water and fat; which both contain mobile, spinning protons. In the absence of a magnetic field, protons in body tissues spin randomly in different directions; however within the strong field of an MRI scanner, these spins *precess* (oscillate like a toy top) around the direction of the strong magnetic field and so most of the body becomes slightly magnetized.

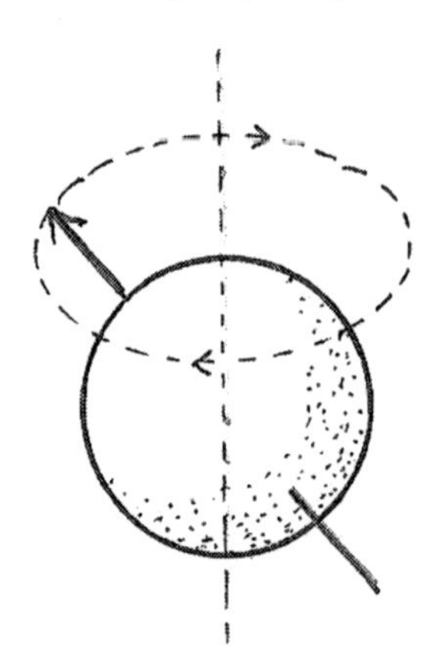

54. Diagram of a precessing proton. The proton has a positive charge so as it spins it creates a magnetic moment which prevents it lining up within a strong **magnetic field** (Bo). Instead it precesses around the axis of the field as well as spinning around its own axis.

As a young man Galileo observed that *"each pendulum has its own time of vibration so definite and determinate that it is not possible to make it move with any other period (frequenza) than that which nature has given it."* A child's swing has to be pushed so as to be in time *(in-phase)* with its natural periodicity to make it swing; any other will stop it. Likewise a proton *precessing* (oscillating) under the influence of a strong magnetic field has its own periodicity at which it will absorb pulses of energy.

This turns out to be at the lower end of the electromagnetic spectrum in the radio frequency range, and its frequency is proportional to the strength of the magnetic field. A proton in a field of precisely one ***tesla*** *precesses* with a frequency of precisely 42.6 MHz . Magnets used in standard MRI have field strengths ranging from 0.2 tesla (9.5 MHz) to 2 tesla (95.2 MHz). These frequencies are in the FM radio spectrum.

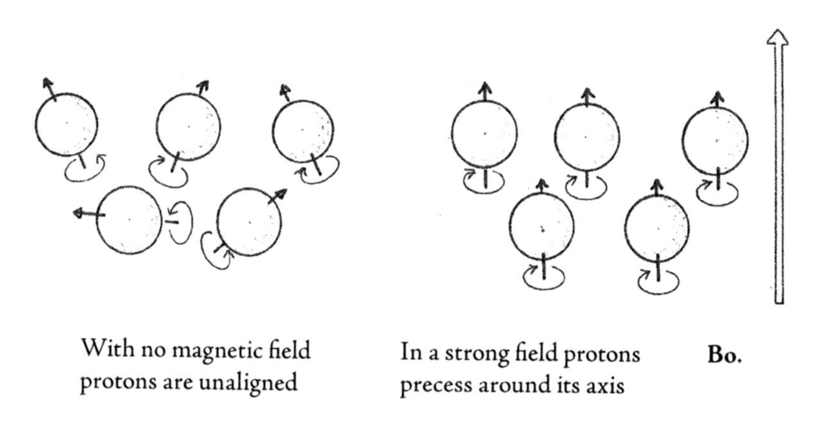

55. Precession of hydrogen nuclei (protons) with no magnetic field and in a strong magnetic field [**Bo**].

The relationship between magnetic field strength and the precessional frequency (radiofrequency) of a proton is simply expressed by the **Larmor equation** [14]. This equation is an absolute law for MRI and its conditions must be met precisely before protons can absorb radio frequency energy; as a child's swing will only swing if pushed (pulsed) in-phase. Once energy has been absorbed by protons within a specific voxel at a specific frequency it is then emitted at exactly the same resonant frequency. Magnetic field gradients are used to individualize voxels by radiofrequency.

Gradient coils alter the magnetic field within the main field to select planar sections within the body and to divide these sections into voxels, The planar sections can be chosen as easily in any orthogonal plane to image axial, sagittal and coronal sections. This is useful in the nervous system to demonstrate parts of the brain and the spinal cord throughout its length.

It is possible using MRI gradient coils, to give every voxel in an anatomical section, its own unique magnetic field, so that according to the Larmor equation, each voxel of tissue can absorb and emit radio energy at its own unique frequency. It is as though every voxel in a body section had its own radio station to receive and return signals.

Voxels can be cubes of tissue with volumes as small as 5-10 cubic mm or even less: sufficient for good spatial resolution of the MR image. A voxel of 10 cubic mm contains approximately 10^{20} hydrogen atoms. In a 512 x 512 matrix in an MRI planar section there are 262,144 voxels, each receiving and transmitting complex radio signals, all of which are detected by radio aerials or coils around the patient. The computer has to sort out all these radio frequency signals to assign data to matching pixels in the image [15].

With an MRI scanner it is possible to map the proton densities of tissues by mapping the amplitude of signals received from the voxels in different anatomical structures. The unique advantage of MR is that it can reflect the different ways in which protons absorb and then emit radio energy. Protons experience differing magnetic forces in tissues with varied physical characteristics and so they resonate in a variety of ways.

Galileo had noticed the "*wonderful phenomenon of the strings of the spinet ...a string that has been struck will begin to vibrate and continues the motion as long as one hears the sound (risonanza). The vibrations cause the immediately surrounding air to vibrate and strike the strings of the same instrument and even those of other instruments... a string tuned in unison with the one plucked is capable of vibrating with the same frequency...after receiving two, three, twenty or more impulses, delivered at proper intervals, it finally accumulates a vibratory motion equal to that of the plucked string.*"

This eloquent description of acoustic resonance and the transfer of pulsed energy from one oscillator to another tuned in unison, are paralleled by the radio frequency ***resonance*** of hydrogen ***nuclei*** in ***magnetic*** fields: **nuclear magnetic resonance.**

Hydrogen protons in different types of tissue water or in lipids live in intrinsically different magnetic environments. In cerebrospinal fluid (CSF) or urine, all the water molecules are free. Their protons have a uniform environment so that they resonate between themselves and retain radio frequency energy much longer than protons in water bound to large protein molecules, or protons in lipids (fat). Free water is said to have a longer **relaxation time** than bound water or lipid, because it resonates longer before returning to the resting state; that is the condition before the excitation radio pulse. It is possible from radio signals to plot the relaxation times of tissues and to map the distribution of differently bound water and lipid protons within the body.

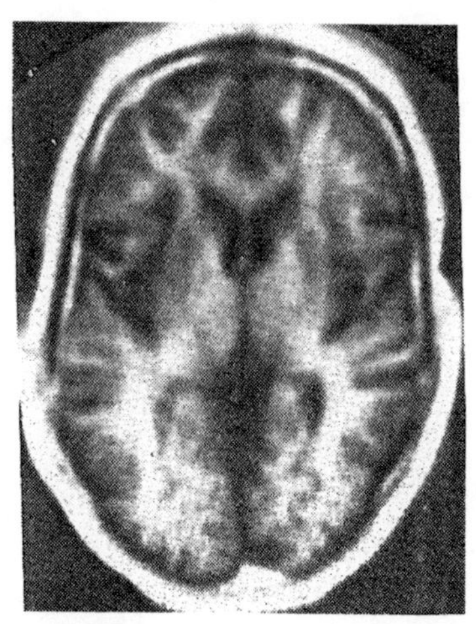

56. T1 relaxation time MR image of normal brain. White matter and scalp fat are bright, grey matter is grey and CSF in ventricles is black.

In the brain the relaxation times for CSF are three times longer than for grey matter and four times longer than for white matter. Signals returning soon after the excitation pulse are greater from white matter than grey matter and much more than CSF so the **T1 relaxation time** brain image shows white matter as white, grey matter as grey and CSF as black. However for late signals this is different because CSF continues to give a signal for much longer and on **T2 relaxation time** images it appears brighter than either grey or white matter. MRI is much more sensitive than x-ray CT at showing many types of brain diseases such as brain tumours, multiple sclerosis, viral encephalitis and degenerative disorders.

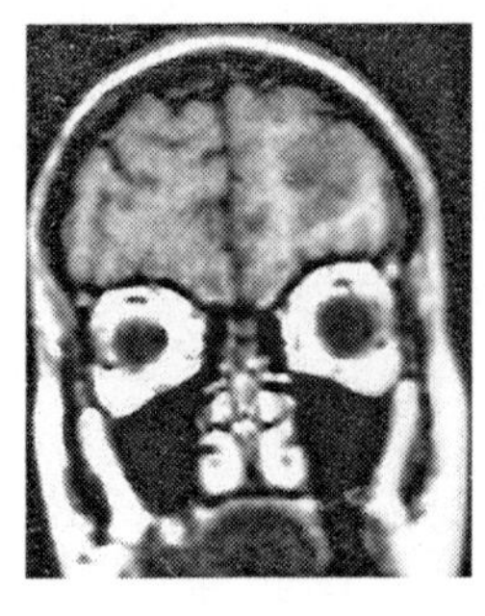

a.

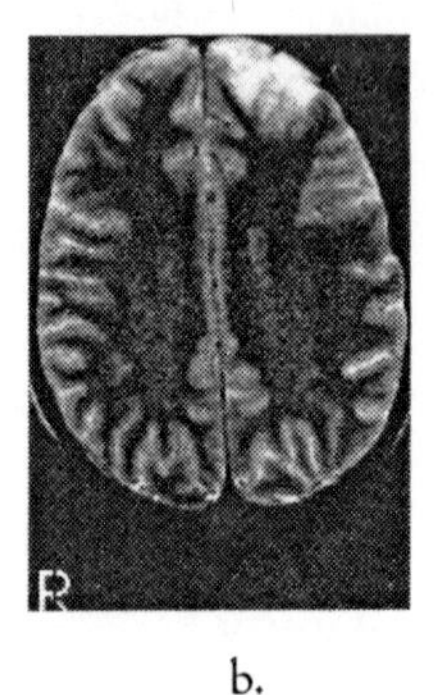

b.

57. MRI Scans of a patient with a left frontal lobe glioma (malignant brain tumour).a. T1 relaxation time image in coronal plane after gadolinium contrast injection shows ring contrast enhancement around the tumour but its necrotic centre remains dark.**b. T2 relaxation time imaging in the axial plane** without contrast shows the tumour as bright.CSF in the lateral ventricles and sulci is also bright.

Elsewhere in the body, inflamed tissues and cancers usually contain larger amounts of free water compared to normal tissues; so they show long relaxation times and appear bright on T2 images. Although compact bone gives no proton MR signals; muscles, tendons, ligaments and bone marrow do. MRI has therefore become useful in showing muscular abnormalities and internal derangements of joints such as the shoulder and knee; and is even more sensitive than radionuclide bone scanning for showing bone tumours. Where MRI is available it has replaced myelography in the diagnosis of prolapsed inter vertebral discs in the cervical and lumbar spine.

Many different new pulse sequences have been developed so as to get different types of information about body tissues. These including very rapid sequences for non-invasive cardiac imaging and for demonstrating flowing blood in arteries and veins. Gadolinium compounds are paramagnetic and can enhance vascular MR images as iodine contrast agents do for x-ray CT.

Oxygen is slightly paramagnetic and so oxy-haemoglobin alters relaxation times of surrounding protons more than deoxy-haemoglobin.

Active areas of the brain have increased blood flow and higher concentrations of oxy-haemoglobin than dormant areas, so that it is possible with MRI to see activation of brain regions in conscious subjects as they carry out various mental tasks.

Blood oxygen level dependent (BOLD) imaging is a method of functional brain imaging, which is swifter and more repeatable than PET. It is now being used as a research tool by many different academic disciplines and is generally known as **functional magnetic resonance imaging (*f*-MRI).**

Other nuclei such as those of carbon-13, fluorine-19 and phosphorus-31 exhibit nuclear magnetic resonance but are not generally used for imaging [16].

The major limitation of MRI, as with CT is its expense. However health administrators and economists are beginning to learn that these instruments can give reliable diagnoses quickly and non-invasively and so reduce overall medical costs. Large medical magnets are expensive to make and operate. Electromagnets use a lot of electrical power, most of which turns to heat. An electromagnetic MRI cooling system can provide sufficient hot water for a small hospital.

Super conducting magnets use liquid helium to cool the magnet coil close to absolute zero, at which temperature electrical resistance ceases, so that a 70 amp current can circulate endlessly to induce a field, so long as liquid helium is topped up. Helium cooled super conducting magnets give much stronger and more uniform fields and therefore better images than electromagnets but are more expensive to make.

There is no risk of ionising radiation with MRI but it has some risks of its own. Not all radio frequency energy is returned as radio signals. A lot is deposited as heat in tissues and can raise their temperature, which may be dangerous for organs such as the brain, spinal cord or the eye. Strong fields pose a theoretical risk of causing cardiac arrhythmias but this is unlikely at the levels used for clinical MRI.

Powerful magnets can exert enormous forces on ferromagnetic objects in their immediate vicinity. Medical scissors and even needles can become missiles in the bore of a magnet. Several anaesthetists have been trapped when their machines have bonded with a magnet. Neuro-surgical clips, especially those on cerebral blood vessels might be twisted by the magnetic field, so patients with these cannot be examined by MRI.

Fortunately orthopaedic metalwork is mostly made of high-grade steel, which is not very magnetic, so MRI is possible in patients with screwed plates or joint replacements.

A final limitation of MRI is that patients must lie in the narrow bore of a powerful magnet and suffer the staccato noises produced by gradient coils as they change magnetic fields. This is a claustrophobic experience and many patients require sedation to endure it.

The amount of anatomical and physiological information that can be obtained with MRI is supreme and well worth the discomfort and risk, which are much less than for invasive angiography or myelography.

Functional imaging allows a neuroscientist lying inside an MRI magnet, to view the activation of different regions of her brain as it thinks about itself. This is a landmark achievement for physics and the exploration of the living body.

COMMENT

Within a century medical imaging has developed from making photographs of finger bones to real-time functional imaging of a thinking brain: from locating embedded bullets to mapping neurotransmitter receptors in the brains of drug addicts. As each medical imaging modality emerged it showed anatomy and physiology in different ways. Many of the twenty thousand diseases to which human flesh is prone were revealed by their impact on the physics of tissue molecules: altered electron densities; uptake of gamma emitting nuclides; changes in echogenicity and radio frequency resonance.

Other modalities may yet develop such as cerebral magnetography for functional brain imaging. In 1965 Gordon Moore, a microchip company executive, made an empirical observation that the processing power of integrated computer circuits doubled every two years. 'Moore's Law' continues to apply in the 21st century. Future developments in imaging will depend on the processing power of computers, which have already contributed to the exponential developments of CT, SPECT, PET and MRI over the last twenty years. Co-registration of information from different modalities has already begun with PET/CT and will progress. Computerized automated pattern recognition might also help to reduce human observer error in diagnostic medical imaging.

All modern medical imaging modalities are available in Europe, North America and the prosperous cities of Asia and South America. But in parts of the world, particularly rural Africa and Asia they are not. Here x-ray sets in rural hospitals are often old and broken. There may be no film, even to x-ray fractures, so bone setting is back in pre-Röntgen days; crippling mal-united fractures are common. Cheap ultrasound machines are more reliable, do not need film and in countries such as Uganda where radiographers have been trained to use them, they have improved obstetrics and helped to locate and drain deep abscesses. Ultrasound works much better in small thin people; CT is ideal for the large and fat.

There has been an automatic assumption, as each new imaging modality appeared that doctors would know what the new images showed; however every modality has given surprises as individual diseases have revealed themselves in a new light. Some landmark imaging revelations were: the ability of chest x-rays to show pulmonary masses, pneumonia and tuberculosis: ultrasound to show foetal abnormalities: gamma imaging to show thyroid function and then the functions of other organs: CT to calculate electron densities of normal and pathological tissues with enormous accuracy: MRI to show small cancers, multiple sclerosis, derangements of joints and brain function invisible to other methods. Every modality in its turn has pushed radiologist physicians up diagnostic learning curves, involving trial and error in the process.

Radiological findings cannot stand alone, but must be integrated with other medical information about the patient: clinical history and examination, endoscopy or pathology, so as to achieve reliable diagnoses for planning effective treatments.

NOTES

1. The **dose of x-rays** is controlled by adjusting the exposure time in seconds and cathode current in milliamps, to give a dose calibrated in **milliamp seconds (mAs)**. The **penetrating power** of the photons or the 'hardness' of the x-ray beam is controlled by the **kilovoltage (kV)**. These factors are adjusted so as to make a satisfactory radiographic image of the part of a patient's body under examination.

2. Technetium-99m is a very useful radionuclide because as ^{99m}Tc pertechnitate it behaves like chloride or iodide ions and can be used for blood flow and thyroid studies

3. Radionuclides and radiopharmaceuticals for gamma imaging.

Radionuclide	Clinical application
Technetium-99m colloid	liver, spleen, red marrow
Tc albumin microsheres	lung perfusion
Tc diphosphonates	bone destruction or repair
Tc pertechnetate	thyroid, brain, cardiac cycle
Tc HIDA	biliary tract
Xenon-133 inhaled gas	lung ventilation
Thallium-201	myocardial perfusion
Iodine-123	thyroid
Indium-111 white cells	abscess location
Gallium-67	abscess and tumour location

4. The SI unit of absorbed radiation dose is the ***gray (Gy)*** and is measured in terms of the energy absorbed per unit mass of tissue; energy in joules and mass in kilograms.

$$1\ Gy = 1\ J\ kg^{-1}\ \text{of tissue}$$

There is an older unit which is sometimes used, the ***rad*** (radiation absorbed dose), which for x-rays, was equivalent to a ***roentgen.***

$$1\ rad = 0.01\ Gy = 0.01\ J\ kg^{-1}\ \text{of tissue}$$

5. For x-rays and gamma rays the biologically effective dose in ***sieverts*** is equivalent to the tissue-absorbed dose in ***grays (Gy)*** or ***1 Joule per kilogram.*** For neutrons the effective dose in sieverts is ten times and for alpha particles is twenty times the dose in grays.

6. Humans are exposed to many sources of ionizing radiation throughout our lives, from cosmic radiation (0.26 *mSv* per year); naturally occurring radiation in our environment such as radon from granite rocks (0.1 to 50 *mSv:* average 1.3 *mSv*); gamma rays from the ground and building materials (0.35 *mSv* per year) and radionuclides within our own bodies, such as potassium-40 (0.30 *mSv* per year).

7. Total natural background radiation averages about 2.2 *mSv* per person per year. Man-made radiations include radioactive fall-out from atmospheric testing of atomic bombs and nuclear accidents, such as Chernobyl (0.02 *mSv* per year), **but the major contribution of man-made radiation to the average citizen in America or Europe is made by diagnostic radiology (0.4 *mSv* per year).**

8. Radiation doses for clinical diagnostic procedures.

Procedure	**Effective dose *mSv* or absorbed dose *mGy***
chest radiograph	0.2
intravenous urography	5
barium meal	5
barium enema	9
CT brain scan	2
CT abdomen scan	8
fluoroscopy	1 per minute
technetium-99m cardiac scan	6
gallium-67 abscess scan	18

9. The acoustic impedance for water is 1.5: for blood is1.59: for muscle is 1.7 and for fat is 1.4 so that ultrasound can pass through interfaces between these types of tissue easily and is partially reflected and transmitted at the tissue boundaries between them.

10. A new type of digital subtraction scanner, called a **synchrotron**, can detect the unique quantum electrodynamics of iodine atoms to overcome motion problems. The method uses the fact that the inner orbital (K) electrons of iodine atoms are bound in such a way that x-rays with photon energies above 33 KeV are needed to displace them. In consequence, photons above this energy are attenuated much more by iodine atoms than photons below 33 KeV. This boundary is known as the **K edge of iodine.**

A synchrotron scans the body with two x-ray beams simultaneously: one beam above and the other below 33 KeV. The lower image is subtracted from the higher energy image to make a synchronous subtraction image of arterial iodine that is not blurred by movement. With a synchrotron it is possible, after a small intravenous injection of contrast, to image the mobile coronary arteries without arterial catheterization.

11. Thousands of measurements were made as the gantry rotated 180° around the patient's head. From these measurements the computer calculated the x-ray attenuation of each and every voxel in an axial section of the head. Hundreds of thousands of differential equations had to be solved by the computer to plot the **attenuation coefficient** of every voxel, using a mathematical technique known as *filtered back projection.* The attenuation coefficient of a voxel represented the electron densities of the atoms within it. These coefficients were assigned arbitrary numbers, which ranged from **+1,000** for compact bone [Z=15] to **–1,000** for air [Z=0], with pure water **0** [Z=7.5] midway between.

12. Attenuation coefficients became known as **Hounsfield units (HU)** and in theory, represent a grey scale with a range of 2,000 units instead of just the four plain radiographic values of bone, water, fat and gas. But the human eye can only distinguish 16 shades of grey so it is not possible for an observer to assess all the information in a CT scan in one view on a TV monitor. Instead it is necessary to interrogate the computer's memory by 'windowing' up and down at different HU levels: below – 600HU to look at air in the lungs: between –10 and + 80HU to look at the brain, heart, abdominal organs or other water-density soft tissues: and above +400HU to look at bone and skeletal anatomy.

13. A positron cannot survive for very long because it soon meets an electron. Both particles are annihilated, releasing energy (according to Einstein's formula: $E = mc^2$) as paired *annihilation photons* with *photon energies* of 0.51 MeV. The two gamma photons are emitted at 180° to one another known as *dual emissions* or *coincident photons* because they are emitted at the same instant from the same point of origin.

A PET scanner has a 360° ring of detectors, which identify coincident 0.51 MeV gamma photons. A powerful computer then maps the locations of annihilations into images, which can be either projections or tomograms in any plane.

14. The relationship between magnetic field strength and the precessional frequency of a proton is simply expressed by the **Larmor equation:** $\omega = \gamma \mathbf{Bo}$, where ω is the **precessional frequency in Hertz,** γ is the **gyromagnetic ratio** for protons, and **Bo** is the **strength of the magnetic field** in **tesla.** The Larmor equation is absolute law for MRI and its conditions must be met precisely before protons can absorb radio frequency energy. Once energy has been absorbed it is then emitted at exactly the same resonant frequency.

15. The transformation of radio frequency signals to spatially related digital data requires special mathematics known as *Fourier analysis* and MRI computers use **two-dimensional Fourier transformations** and sometimes **three-dimensional Fourier transforms** to generate tomographic images.

16. However ^{31}P signals can be superimposed on proton MR images of the brain, to monitor metabolic changes in injured tissues. The relative amounts of ATP, ADP and inorganic phosphates vary in healthy, dead or dying tissues and can be assessed with ^{31}P NMR to indicate the viability of tissues. MR assessment of phospho-energetic processes can be used to monitor the progress of infarcted brain tissue in stroke patients.

CONCLUSION

Two important themes recur in the stories of discovery in this book. The first is the close relationship between the physical and medical sciences; the second is the inspirational nature of discovery and the power of inventive genius to formulate surprising theories of great explanatory and predictive power; theories that revolutionized scientific ways of looking at the natural world.

The discoveries emphasize that the laws of physics govern the living human body as they do inanimate matter. For most ordinary people this unsurprising fact has been made difficult to accept; mainly because physics and biology have been publicly presented as separate disciplines.

The physics of quantum electrodynamics helped to understand and explain DNA and human genetics. The Human Genome Project completed in 2003 resulted from the discoveries of physicists as well as medical scientists and promises to deliver further insights into our nature. Quantum and radiation physics have provided new technologies such as ultrasound, nuclear medicine, CT and MRI for the non-surgical exploration of the living body.

By digitizing anatomical information so as to provide quantitative data about electron density of tissues or by accurately locating the nucleotide bases in strands of DNA, computers have brought human and molecular biology within the ambit of the '*exact sciences*'. William Whewell, philosopher of science and Master of Trinity College,

Cambridge would have rejoiced that the three distinct sciences he defined in 1833; *physics, chemistry and biology*, have united. It can now be accepted that the separation between *physics* and *human biology* is artificial.

The laws of physics apply at every level to the ions, molecules, cells, tissues and organs of our bodies; physics goes on inside us as well as outside.

As a fellow of Trinity, Whewell focused on the scientific work of two previous fellows; Francis Bacon and Isaac Newton. He realized that Bacon's account of scientific induction and discovery was too mechanical and did not really represent the thought processes by which great discoveries had been made. Whewell came to view scientific discovery as an imaginative and creative process; he recorded Newton's *"extreme absence of mind"* and *"transcendent powers"*. Of Newton he wrote *"he knew not what he did, and his mind appeared to have quite forgotten its connexion with the body..."*

Determination and inspired intuition by prepared minds seem to be common factors for all the great advances in theoretical and experimental science since Galileo and Harvey.

Although the major theme of this book is that the laws of physics apply to living bodies, especially our own, this does not mean that I support the argument that everything can be explained by physics alone. The proposition that conscious awareness is just an *epiphenomenon* resulting from physical processes in the brain may be a way of disposing of the *'consciousness problem'*, but it is not a scientific proposition and has no explanatory power. Although a conscious human mind requires a functioning brain, physics cannot explain how it feels to be alive.

Subjective sensation of red = electromagnetic wavelength of 600 nanometres does not compute. A neuroscientist in a functional brain scanner can view his or her own functioning brain in real-time, thinking about viewing their own functioning brain thinking..., but may never able to formulate a mathematical statement about subjective awareness of self or the mystery of consciousness.

Great mathematicians and natural philosophers; Descartes, Pascal, Newton, Helmholtz, Bohr, Schrödinger and Penrose, have each tackled the mind-matter problem without mathematical success. Physics has this limitation; it can explain much of what happens in the living human body, but not much about the mind. The reason is simple; mathematical physics is a creation of the human mind and therefore an adjunct to it. An observer cannot observe itself objectively. Only artists and poets can attempt to describe what it feels like to be alive.

"It ought to be generally known that the source of our pleasure, merriment, laughter and amusement, as of our grief, pain anxiety and tears is none other than the brain. It is specifically the organ that enables us to think, see and hear; to distinguish the ugly and the beautiful, the bad and the good, pleasant and unpleasant. Sometimes we judge according to convention and at other times according to expediency. It is the brain too which is the seat of madness and delirium, of the fears and frights which assail us, often by night, but sometimes even by day, of thoughts that will not come, forgotten duties and eccentricities."

These ideas were expressed in the *Hippocratic Corpus* 2,500 years ago and are the thoughts of a great physician.

READING

PHILOSOPHERS PHYSICISTS AND PHYSICIANS, & LIVING MATTER

Chadwick J & Mann WN	'The Medical Works of Hippocrates'	Blackwell 1950
Crump T	'The Mastery of Fire' and 'Religion & the Soul' in a 'Brief History of Science'	Robinson 2000
Freeman C	'The Closing of the Western Mind – The Rise of Faith & the Fall of Reason'	Heinemann 2002
Franklin B	'The Physics of Aristotle' in 'The Works of Aristotle I' by Thomas Taylor in 1806.	Promethius Trust reprinted 2000.
Kelly EC	'The Theory and Practice of Medicine By Hippocrates.'	Citadel Press 1964
Morson J	'Theory and Practice in Science' In 'Culture and Science in China'	ABC Sydney 1981
Porter R	'Medicine and the Body' in 'Flesh in the Age of Reason'	Penguin 2003
Quentin A	'Introduction' to 'Plato's Symposium' Translated by T. Griffiths	Folio Society1991

Quentin A 'Religion and Science' in 'From Wodehouse to Wittgenstein' Carcanet Press 1998

Vogel S 'Connecting up Muscle' in 'Prime Mover, a History of Muscle.' Norton & Co 2001

Wood M 'In the Footsteps of Alexander the Great' BBC Books 1997

TIME & MOTION, FLESH & BLOOD

ITALY

Chamberlin ER 'The World of the Italian Renaissance' Book Club 1982

De la Croix H 'Proto-Renaissance Italy' & '15th century' In 'Art Throughout the Ages' HBJ 6th Ed 1975

Garcia-Ballester L 'Galen and Galenism' Ashgate Varorium 2002

Hawking S 'Galileo Galilei – His Life and Work' and 'Dialogue Concerning Two Sciences' in 'On the Shoulders of Giants' Penguin 2002

Hibbert C 'Venice – 'The Biography of a City' Grafton 1988

Hunt FV 'Origins in Acoustics' Yale UP 1978

Sobel D 'Galileo's Daughter – a Drama of Science, Faith and Love' Fourth Estate 1999

Vogel S 'and how we found out' in 'Prime Mover – A Natural History of Muscle.' Norton & Co 2001

Sharratt M 'Galileo: Decisive Innovator.' Cambridge UP 1994

THE NETHERLANDS AND ENGLAND

Anscombe E & Geach T 'Descartes Philosophical Writings' Nelson 1975

Ashley M	'The English Civil War'	Sutton 1990
Aughton P	'Newton's Apple, Isaac Newton and the English Scientific Renaissance'	Weidenfeld & Nicolson
Crump T	'A Brief History of Science'	Robinson 2000
Gleick J	'Isaac Newton'	Harper 2004
Hay D	'A History of the Sydenham Medical Club'	Hay 2001
Hawking SW	'A Brief History of Time'	Bantam 1990
Hawking SW	'Isaac Newton' and 'Principia' in 'On the Shoulders of Giants'	Penguin 2002
Hughes JT	'Thomas Willis, 1621 –75, His Life and Work'	Royal Society of Medicine
Jardine L	'The Curious Life of Robert Hooke, The Man who measured London'	Harper Collins 2003
Jardine L	'Britain and the Rise of Science' In 'Revolutions in Science'	BBC Science and Discovery 2002
Keele KD	'William Harvey: the man, the physician and the scientist.'	Nelson 1965
Locke J	'An Essay Concerning Human Understanding'	Penguin Classics
Lomas R	'The Invisible College'	Headline 2002
Porter R	'Flesh in the Age of Reason'	Penguin 2003
Shakelford J	'William Harvey & the Mechanics of the Heart.'	Oxford UP 2003

Tomalin C — 'Samuel Pepys: the Unequalled Self' — Penguin 2002

HUMAN ENERGY

Allan DGC & Schofield RE — 'Stephen Hales: Scientist & Philanthropist' — Scholar Press 1980

Cahan D — 'Mechanical Foundations of Thermodynamics' by Bierhalter G
'Studies in Physiological Heat' by Olesko K & Holmes FL
in 'Hermann Helholtz & the Foundations of 19th C Science'
editor Cahan D — California Univ Press 1993

Crump T — 'Energy: Science Refounded' &
'Chemistry: Matter and Transformation'
In 'A Brief History of Science.' — Robinson 2001

Gibbs FW — 'Joseph Priestly: Adventurer in Science
and Champion of Truth.' — Nelson 1965

Grande F & Visschler MB — 'The Discovery of Glucose and Glycogen'
& 'The Concept of Mileu Interior' in Schenkmann
'Claude Bernarde and Experimental Medicine.' — Publishing 1967

Guerlac H — 'Antoine-Laurent Lavoisier:
Chemist and Revolutionary.' — Chas. Scribner's Sons NY 1975

Ramsay W — 'The Life and Letters of Joseph Black' — Constable 1918

Uglow J — 'Experiments with Air', 'Fire' and 'Steam' in
'The Lunar Men: The Friends who made the Future'
Faber & Faber 2002

Vogel S — 'Body Work' and 'Working Hard'
in 'Prime Mover-A Natural History of Muscle' — Norton &Co 2001

ELECTRICAL EXCITEMENT

Broks P — 'Articles of Faith' in 'Into the Silent Land' — Atlantic 2003

Burrell B	'The Most Complex Object' & 'Einstein' in 'Postcards from the Brain Museum'	Broadway Books 2004
Clarke E & E	'The Neuron','Nerve Function','The Reflex'in 'The Human Brain and Spinal Cord. A History'	California Univ Press 1968
Crump T	'Discovering Electricity' in 'A Brief History of Science.'	Robinson 2001
Erlanger J	'Some Observations on the Responses of Single Nerve Fibres' Nobel Lecture 1947	Elsevier 1964
Feldberg W	'The Early History of Synaptic and Neuro-Muscular Transmission by Acetyl-choline' *and*	
Hodgkin AL	'Chance and Design in Electrophysiology' *and*	
Huxley AF	'Looking Back on Muscle' in 'The Pursuit of Nature. Essays on the History of Physiology'	Cambridge UP 1977
Grande F &	'Physiology of Junctional Transmission'	Schenkmann Pub Co
Visscher MB	'Claude Bernard & Experimental Medicine'	Cambridge, Mass.
Greenfield S	'The Private Life of the Brain.'	Penguin Books 2000
O'Connor WJ	'Founders of British Physiology'	Manchester UP1988
Olesko K & Holmes FL	'Early Researches – muscle & nerve physiology' in 'Hermann Helmholtz & the Foundation of 19th C Science' Ed Cahan D.	California UP 1993
Pancaldi G	'The Battery' in 'Volta'	Princeton Press 2003
Penrose R	'We need new physics to understand the mind' in 'Shadows of the Mind'	Vintage 1995

Poore V	'Electricity' in 'Quaine's Medical Dictionary'	London 1886
Rapport R	'Nerve Endings, the Discovery of the Synapse. The Quest to find how Brain Cells Communicate.'	Norton Press 2005
Ryle G	'Sensation & Observation' in 'The Concept of Mind'	Penguin 1973

VISIBLE LIGHT

Anscombe E & Geach PT	'The Visible World' & 'Dioptrics' in 'Descartes Philosophical Writings'	Thomas Nelson 1970
Broks P	'Einstein's Brain' in 'Into the Silent Land'	Atlantic 2003
Calder N	'Chapters 2 & 5' in 'Einstein's Universe'	Penguin 1983
Crump T	'The Rebirth of Science' in 'A Brief History of Science'	Robinson 2001
Feynman RP	'QED the Strange Story of Light and Matter'	Penguin 1985
Gleick J	'Isaac Newton'	Harper 2004
Gregory RL	'Eye and Brain'	Wiedenfeld & Nicholson 1972
Gribbin J	'Photons and Electrons' in 'In Search of Schrödinger's Cat'	Black Swan 1991
Hawking S	'Space and Time' & 'The Unification of Physics' In 'A Brief History of Time.'	Bantam Press 1988
Hawking S	'On the Electrodynamics of Moving Bodies – Einstein' 'On the Shoulders of Giants'	Penguin Books 2002
Hendry J	'James Maxwell & the Theory of the Electromagnetic Field'	Adam Hilger 1986

Hunt BJ — 'The Maxwellians' — Cornell Univ Press 1991

Segne E — 'What is Light?'& 'Thomas Young a Universal Talent'
'From Falling Bodies to Radio-waves' — WH Freeman & Co 1984

Tolstoy I — 'James Clerk Maxwell' — Canongate Edinburgh 1981

Turner RS — 'Visual Perception of Space' &
'The Eye as Mathematician' in
'Hermann Helmholtz & the Foundations of 19th C Science'
Editor Cahan D. — Calif Univ Press 1993

NEW RAYS, NEW ATOMS AND LIVING MOLECULES, & RADIOLOGY - EXPORING INNER SPACE

Baddeley H — 'Radiological Investigation –
A Guide to Medical Imaging in Clinical Practice' — John Wiley 1984

Campbell J — 'Rutherford – Scientist Supreme.'

Davies P &
Gribbin J — 'Quantum Wierdness' and 'The Living Universe'
in 'The Matter Myth, beyond Chaos & Complexity' — Penguin 1992

Dawkins R — 'Puncturing Punctuationism' in
'The Blind Watchmaker' — Penguin 1988

Farr RF &
Allisy Roberts P — 'Radiation Physics' in
'Physics for Medical Imaging' — Saunders 1998

Feynman RP — 'QED The Strange Theory of Light and Matter' — Penguin 1990

Glasser O — 'Wilhelm Conrad Röntgen and the Early History of Roentgen Rays' — Norman Publish. 1993

Grainger RG &
Allison DJ — 'Diagnostic Radiology' 2nd Ed — Churchill Livingstone 1993

Hawking SW	'Galileo' in 'On the Shoulders of Giants'	Penguin 2002
Gribbin J	'In Search of Schrödinger's Cat'	Black Swan 1984
Holtzmann B	'Discovery of X-rays', 'The Other Shoe' in 'Naked to the Bone: Medical Imaging in the 20th Century'	Helix 1997
Mould RF	'Discovery of X-rays, Radioactivity and Radium' 'Military Radiography', 'Radiation Risks' in 'A Century of X-rays and Radioactivity in Medicine'	Institute of Physics Publishing 1993
Schrödinger E	'What is Life?' & 'Autobiographical Sketches'	Cambridge UP 1992
Snow CP	'Founding fathers', 'The Quiet Dane', 'The Golden Age','This will never happen' in 'The Physicists'	Macmillan 1981
Watson JD	'The Double Helix'	Penguin 1968
Yeo R	'William Whewell: Cambridge Historian and Philosopher of Science' in 'Cambridge Scientific Minds' Editors Harman P & Mitton S	Cambridge UP 2002

WEBSITES

American Institute of Physics	www.aip.org
Institut Curie	www.curie.fr
Council for the Central Laboratory of Research Councils	www.cclrc.ac.uk
John Dalton	//rylibweb.man.uk/Dalton
Wilhelm Conrad Röntgen	//rontgen.nobelpr.com
Ernest Rutherford	www.rutherford.org.nz
The Frederick Soddy Trust	www.soddy.org

GENERAL REFERENCE

American Academies Of Science	'The Dictionary of Scientific Biography.'	Chas Scribner's Sons NY 1977
BBC Science	www.bbc.co.uk/ Science & Technology	
Brown BH & Smallwood RH	'Medical Physics and Biomedical Engineering'	Institute of Physics 1999
Bynum WF & Porter R	'Companion Encyclopaedia of the History of Medicine'	Routledge 1993
Fraser G	New Physics for the Twenty-first Century	Cambridge UP 2006
Porter R	'The Cambridge Illustrated History of Medicine'	Cambridge UP 1996
Sebastian A	'A Dictionary of the History of Medicine'	Parthenon 1999
	'Encyclopedia Britannica'	University of Chicago.
	'The New Shorter Oxford English Dictionary'	Clarendon Press.

INDEX

Printed in the United Kingdom
by Lightning Source UK Ltd.
135108UK00001B/169-300/P

9 781438 917030